Ever Green

Ever Green

SAVING BIG FORESTS TO SAVE THE PLANET

John W. Reid *and* Thomas E. Lovejoy

W. W. NORTON & COMPANY
Independent Publishers Since 1923

FRONTISPIECE: Strangler fig wraps a host tree in the western Congo rainforest.

All photographs by John Reid unless otherwise indicated.

Maps by David Atkinson. Digital map information prepared by Jessica Reid with guidance from Peter Potapov, and data on 2010 forest cover and 2016 intact forest landscapes provided by the Global Land Analysis and Discovery laboratory at the University of Maryland.

Printed in the United States of America
First Edition

For information about permission to reproduce selections from this book, write to Permissions, W. W. Norton & Company, Inc., 500 Fifth Avenue, New York, NY 10110

For information about special discounts for bulk purchases, please contact W. W. Norton Special Sales at specialsales@wwnorton.com or 800-233-4830

Manufacturing by Lake Book Manufacturing
Book design by Chris Welch
Production manager: Lauren Abbate

ISBN 978-1-324-00603-9

W. W. Norton & Company, Inc., 500 Fifth Avenue, New York, N.Y. 10110
www.wwnorton.com

W. W. Norton & Company Ltd., 15 Carlisle Street, London W1D 3BS

1 2 3 4 5 6 7 8 9 0

John: For Carol, Jessica, Charlie, Mom, and Dad

Tom: For Betsy, Kata, and Annie

CONTENTS

LIST OF MAPS

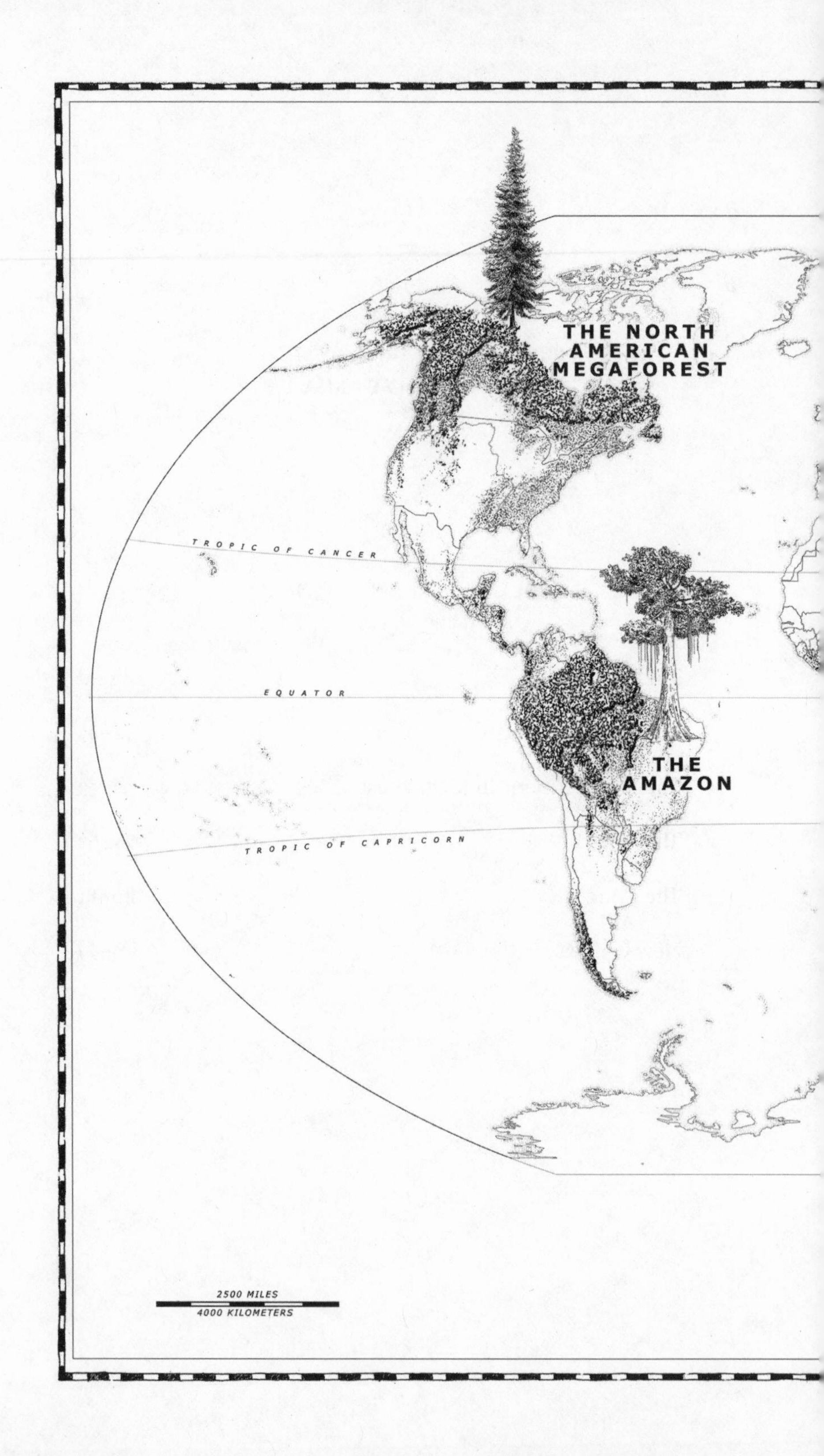
THE NORTH AMERICAN MEGAFOREST
TROPIC OF CANCER
EQUATOR
THE AMAZON
TROPIC OF CAPRICORN
2500 MILES
4000 KILOMETERS

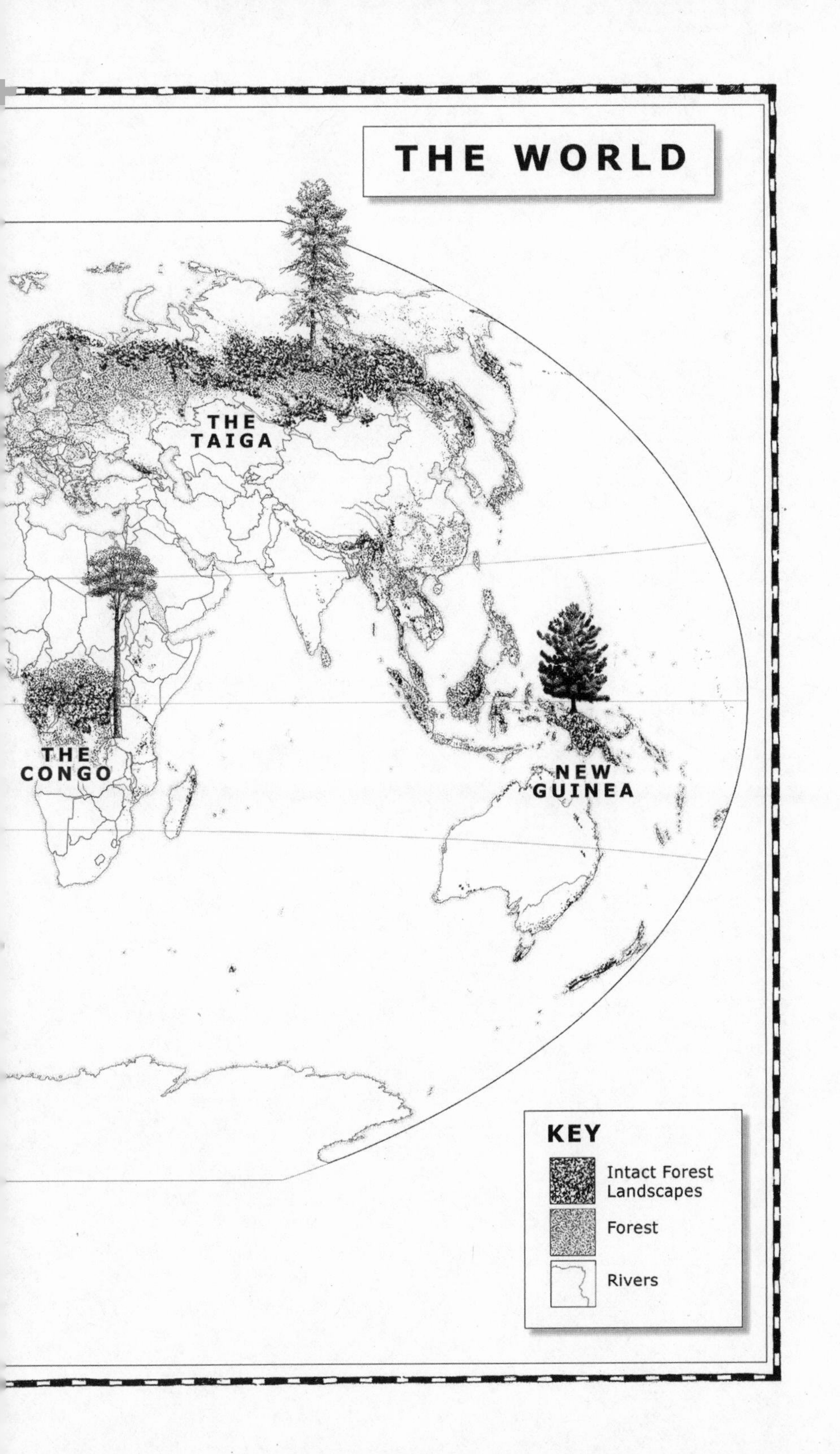
THE WORLD
THE TAIGA
THE CONGO
NEW GUINEA
KEY
Intact Forest Landscapes
Forest
Rivers

A NOTE ON WEIGHTS AND MEASURES

Units of measurement in general use in the United States differ from those in most other countries. Our inches are divided into fractions rather than decimals and are gathered in groups of twelve to produce feet. We have acres for area (43,560 square feet) and miles for distance (5,280 linear feet). The United States has similarly idiosyncratic volume, mass, and temperature measurements. The purpose of this book is not to promote our quirky weights and measures, nor to advocate for the metric system. We wish for the book to be readily understood by readers in our home country and by readers in other parts of the world. With that goal in mind, here is how we've handled the various dimensions and quantities needed to describe big forests and their contents.

General

In general, we emphasize the US system, with metric conversions for the first mention of the units—repeated later in the text where it seems like a good time for a refresher. In some instances, where we are describing scientific findings, for instance, we emphasize the metric measures, which are used in all science, and give their US conversions.

Area

We use acres extensively to describe areas of forest. The acre is roughly half a professional soccer field. 2.47 acres are contained within a hectare, which is the most commonly used metric unit employed by conservationists and scientists for forest areas. Square kilometers, each equal to 100 hectares, are in common use outside the United States to describe especially large areas. The US analogue is the square mile, which is 2.59 square kilometers, or 160 acres.

Distance

We use miles, each of which is equal to 1.61 kilometers.

Volume

US gallons (3.79 liters) are used for liquids. For wood, the US unit is the board foot, each of which contains 144 cubic inches (for instance, a square board 12 inches on a side and 1 inch thick). 424 board feet fit in 1 cubic meter, the metric measure of wood volume.

Weight

The masses of carbon and carbon dioxide are given in metric tons, the unit used in virtually all discussions of the climate-warming gas and the solid element that is part of it. Carbon makes up roughly 50 percent of plant material, also known as biomass, and we also use metric tons for that. When talking about the mass of things other than carbon, we use pounds (2.2 per kilogram) and short tons, which, at 2,000 pounds, are 10 percent lighter than metric tons.

Temperature

Temperature features in our book mainly as degrees of global warming. The measure in degrees Celsius is used globally, including in the United States, and is therefore the pick for our text.

Ever Green

Into the forest in New Guinea's Tambrauw Mountains.

PROLOGUE

Anastasia's Woods

Birds of paradise and crossed hatchets decorate her yellow party dress. Bare feet pick a path over roots and rocks and quilts of moist leaves. Little brown hands grip ferns for balance as she lowers herself down a slope so steep you can reach sideways and touch the earth. She weaves through gaps in shoulder-high limestone, ancient corals that were compressed until they had nowhere to go but up and into a new career as mountains. The girl's feet seem scarcely to feel the brittle edges. She turns her close-cropped head and gives us a smile, then disappears down the path.

Anastasia is a 2-year-old member of the Momo clan, which, for generations beyond reckoning, has lived in this forest in western New Guinea. She is accompanied by her mother, Sopiana Yesnath, a family friend named Mariana Hae, and Anastasia's aunt, Fince Momo, who is shadowed by a limping pointy-eared hunting dog named Hunter. Several visitors scramble to keep up. The way leads into leafy gullies and along a toothy forested ridge. After 3 hours we arrive at a flat spot just big enough for our tents and a fire. Nearby, a clear stream flows over a bed of limestone bulbs formed by the calcium carbonate of ancient shells and exoskeletons that dissolved farther up the mountain.

We set up camp while the elder Momo and her dog head out to fish and hunt. Then Hae takes us for a walk. She points out a tidied bit of terrain belonging to a magnificent bird of paradise. The male of this species displays mating fitness by beak-flinging forest debris away until bare ground is revealed. His two tail feathers are green hoops. He has a blue beak and blue feet, a bright green breast that expands like a cobra's hood, and a yellow-and-brown back. The bird has some red on him, too, and, in an arresting final touch, lime green skin on the inside of his mouth. We follow Hae down a near-precipice, gripping trees. She descends like liquid, only slightly slower than free fall. Near the bottom we consider a loaded langsat tree. The fruits are kiwi-sized, with delicious, tart, translucent white flesh protected by thin leathery skin. But they're out of reach. Hae spiders up into the tree and drops down bunches. When we reach the narrow valley floor, we strip and flop into the emerald current of the Iri River.

This is the heart of a megaforest, one of five stunningly large, wooded territories that remain on Earth. New Guinea is the smallest of the five. It's an island situated just north of Australia, twice the size of California and almost completely covered in trees. Its western half is ruled by Indonesia, while the east is the independent nation of Papua New Guinea.

Next in size is the Congo, occupying Africa's wet equatorial middle, including parts of the Democratic Republic of the Congo, the much smaller Republic of the Congo, Cameroon, Gabon, the Central African Republic, and Equatorial Guinea.

The Amazon is the largest tropical megaforest, roughly double the Congo's extent. It covers most of South America's bulge and is shared by eight independent states—Brazil, Peru, Colombia, Bolivia, Ecuador, Venezuela, Guyana, and Suriname—and the department of French Guiana.

The far north holds the two largest forests on Earth. They are called boreal, after Boreas, Greek god of the north wind. Their boundaries are defined by a mean temperature range of 50°F–68°F (10°C–20°C) in the warmest month of the year. The North American boreal zone starts on Alaska's Bering seacoast, marches across the state, and sweeps southeast through Canada, all the way to its Atlantic shore.

The other boreal forest and largest of all megaforests is called the Taiga.

Mariana Hae on the bank of the Iri River in West Papua.

It's almost entirely in Russia, extending from the Pacific Ocean across all of Asia and far-northern Europe and from the Arctic Circle south to Central Asia.

The megaforests' most unbroken cores are called "intact forest landscapes." This phrase was coined in the late 1990s by a group of Russian scientists and activists to describe the forests that were their top priorities to defend from industrial logging. Timber companies were advancing quickly into Russia's old growth as the country's economy opened to the West in the aftermath of Soviet collapse. The environmentalists came up with a precise definition and a map. The adjective "intact" is earned by encompassing at least 500 square kilometers—which is roughly 125,000 acres—free of roads, power lines, mines, cities, and industrial farms. That's the size of about 60,000 soccer fields, 146 Central Parks, or a single square of land 14 miles on a side. "Landscapes" is added to the term because natural forests have vital treeless places, such as rivers, lakes, wetlands, and mountaintops mixed in. In 2008, the group helped map all such forests globally. Worldwide, there are currently around 2,000 intact forest landscapes, or IFLs, comprising nearly a quarter of all the planet's wooded lands. They are heavily concentrated in the five megaforests.

Our planet needs its megaforests and their IFLs to keep functioning. Global average temperatures have already risen by 1°C (1.8°F) since pre-industrial times. "Once-in-a-century" fires, droughts, floods, and storms are occurring annually. In 2020, Australia had its hottest night, California burned twice as many acres as ever before, and the unprecedented Atlantic hurricane season required the use of both Latin and Greek alphabets to name all the storms. People were starving in desiccated Madagascar, unprecedented expanses of corals were poaching in overheated seas around the Great Barrier Reef, and permafrost was heaving in the north. The climate crisis has left the realm of theory and speculation. It is here.

The Intergovernmental Panel on Climate Change (IPCC) says we need to stabilize warming at no more than 1.5°C (2.7°F) to avert hot social crises and ecological disasters in the future. The climate solutions that we hear about most often, like swearing off coal or switching to electric cars, address the problem by disrupting the industrial processes by which fuels coming out of the ground get into the air. These strategies are absolutely necessary, but they skip what is between rock and atmosphere: the biosphere. The math of keeping our world livable doesn't add up without caring for our planet's biology in general and keeping our big forests in particular. The IPCC finds that all pathways for limiting warming to 1.5°C involve reversing deforestation by 2030.

Over the planet's lifetime, carbon has moved in dramatic volumes between four realms—the atmosphere, oceans, subterranean spaces, and the layer of animate beings. Plants photosynthesize, turning atmospheric carbon into biomass. When conditions are right, in swamps, for example, undecomposed plant matter accumulates and gets packed into coal beds. Oil and gas form thanks to shallow oceans where masses of tiny plants and animals die, get buried, and then get compressed.

Twice before, plants have decarbonized the atmosphere on a massive scale. The first time was around 400 million years ago, when, together with fungi, they expanded onto dry land. Plants really got going when they developed vascular systems, which allowed them to move water around internally and colonize drier environments. Atmospheric carbon dioxide (CO_2) dropped from thousands of parts per million to several hundred. Then, 252 million years ago, a cluster of volcanoes in Siberia erupted, elevating CO_2 in the atmosphere, heating the planet, changing the ocean's chemistry, and killing off most species on both land and sea.

The biosphere gradually recovered, creating new species from the survivors. Around 100 million years ago, something came along that would eventually make humanity possible: flowers. A new group of terrestrial plants that used flowers for reproduction emerged and supplanted conifers as the dominant vegetation over much of Earth. They did it by downsizing their genomes, which meant they could have smaller cells and, as a consequence, pack more veins and carbon-absorbing pores onto each leaf. They grew like crazy and drained carbon from the atmosphere until it reached its current levels, conditions in which humans and the rest of the current biota have thrived. Today's flowering plants include maples, mahoganies, and roses, along with 300,000 other species.

In the blink of an eye, geologically speaking, industry and agriculture are refilling the sky and carbonating the oceans. Human societies need to reorganize production to leave as much carbon underground as possible. We also need to leave it in, and restore some to, the biosphere. The ecosystems densest in carbon are forests, and of these, the most carbon rich are those least disturbed. In the tropics, unfragmented forests hold double the average carbon for all tropical forests. They are wetter, more luxuriant, less fire prone, and fuller of plant matter than jungles pierced by roads and hemmed in by farms. When it comes to underground plant carbon, our planet's biggest deposits lie in the deep soils and peat layers beneath the intact boreal forests. The boreal holds 1.8 trillion metric tons of carbon, 190 years' worth of global emissions at 2019 levels.

Saving big quantities of carbon in intact forests is cheap because these lands are remote and the process is simple. Keeping carbon in tropical forests costs a fifth as much as reducing emissions from energy and industry in the United States or Europe. And it's more affordable by a factor of at least seven than regrowing forests once they've been felled. This opportunity is still, astonishingly, overlooked and unmentioned in most national climate plans.

The megaforests would be remarkable even if all they did was store massive amounts of carbon in featureless mats. Instead, under leafy and needled canopies, these ecosystems have tigers, bears, and 10-pound harpy eagles with fanlike crests. Megaforests have almost all the birds of paradise, as well as giant otters, anacondas, chimpanzees, bonobos, and gorillas. Most of the planet's bugs, trees, mushrooms, and freshwater supplies are in the big woods, as are hallucinogens, analgesics, tumor shrinkers, stomach settlers, anesthetics, vision enhancers, sedatives, stimulants, and more. Life is in full riot in the intact forests. They are the planet's wildest, most biologically diverse lands. In the north, big, iconic creatures like grizzly bears, wolves, cats, caribou, and salmon depend on—and sustain—the woods, as do 3 billion migratory songbirds and waterfowl from the tropical and temperate latitudes. Tropical rainforests, the most diverse of all, regularly produce discoveries of life-forms previously unknown to science. And they are not all tiny things only a biologist could love; twenty new monkeys have been found in Brazil since 2000, three in 2019 alone.

A paradise kingfisher shares territory with the Momo clan in West Papua.

The diversity of people is similarly spectacular in the megaforests. Around a quarter of the planet's roughly 7,000 living languages are spoken in the five great wooded regions. The Amazon has over 350 known languages, plus some never heard outside the forest because they are spoken by uncontacted tribes. The variety of Amazonian grammars has astonished linguists and showcased the boundless inventiveness of the human mind in the art of communication. In the Congo, Pygmy societies remain as forest specialists and spiritual intermediaries even after thousands of years of contact with farmer-neighbors. Russia has Indigenous cultures who trace their ancestry to tigers. In Canada and Alaska, dozens of Native cultures sustain old ties to the severe forest landscapes of the boreal. And the island forest of New Guinea is, by a country mile, the most linguistically diverse place on Earth, with over a thousand languages in an area less than a tenth the size of the United States. Megaforests have offered people socioecological niches in which to differentiate and stay different, buffered from the homogenizing traffic of global ideas and colonial languages.

Over 10 percent of intact forest landscapes were fragmented or lost between 2000 and 2016 (the latest year for which comprehensive data are available). In the far northern boreal forest, mines, oil, and gas, with their seismic lines, roads, and pipelines, are primary threats. In the southern boreal, where trees are bigger and closer to mills, logging carves into many intact areas. All across the boreal zone, fires rage during the short northern summers, hotter and more frequently than in the past. In the tropics, logging and road building are scourges of intact forests. Roads give hunters easier access to deep jungle redoubts and make it feasible to farm previously remote wilderness, particularly in the vast flatlands of Amazonia. As in the boreal, climate shifts and human pressure are bringing more blazes and ecological upheaval to the equatorial woods.

To keep Earth livable, we humans must grow out of our habit of transforming woods into landscapes of grass, shrubs, dirt, and pavement. To start, countries can support forest peoples who steward vast areas of the megaforests. Indigenous peoples, whose cultures, spirituality, and practical survival are tied to the trees, control around a third of intact forests. They're the ones who know the woods best. Fince Momo demonstrated this as we walked in her West Papuan homeland. She pointed out plants used for mattresses, pot holders, roofing, string bags, traditional clothing, arrow shafts, and spears; for treating coughs, stomach trouble, and malaria; for seasoning pork and starting fires; and even one that, when soaked in water, produces an elixir that makes dogs better hunters. She drew our attention especially to the inconspicuous, smallish *kapeswani* leaf. It is the symbol for her clan. In Maybrat, her language, the name means ghost, a reference to specters' bloodsucking proclivity. The leaf is used as a coagulant during caesarean home births. It also makes a nice red dye. Fince Momo vows that her clan will defend the forest at all costs.

On the island of New Guinea, 90 percent of the forest is held by Native communities. Across the Pacific, Brazil's 1988 constitution established the country as a world leader in the legal recognition of Indigenous peoples' rights to their ancestral territories. Neighboring Colombia enshrined similarly strong protections. Along with other Amazon basin countries, they have recognized hundreds of millions of acres of aboriginal forests. The carbon stored in these lands has proved dramatically less likely to be emitted to the atmosphere than anywhere else in the Amazon. In Canada, Indigenous peoples are reasserting control over their ancestral territories, largely with the support of the government, which recognizes the original peoples as teachers and partners in taking care of nature.

Reinforcing these trends is a practical and ethical way to save the megaforests. Indigenous control should expand to encompass full traditional territories rather than being limited to small village titles. Lands should be legally inalienable from their human communities. And traditional forms of territorial control should be respected, not supplanted by simplified ownership.

On other lands, outside of Indigenous territories, protected areas are a proven solution. In fact, they are arguably the greatest environmental success story in the history of modern nations. In 1990, 4 percent of Earth's land was protected. Over the last three decades, governments have collectively quadrupled this figure to 17 percent. Most nations are agreeing to nearly double that global total to protect 30 percent of all land by 2030.

Modern parks emerged in North America with the establishment of Yosemite and Yellowstone, in 1864 and 1872, respectively, and really picked up steam in Russia and the United States around the turn of the twentieth century, propelled by Russian scientists and romantic American outdoorsmen like John Muir and Teddy Roosevelt. In recent years, governments have protected hundreds of millions of acres across the Amazon and Congo basin megaforests. Well-run funds have been set up to enable international contributions toward the modest costs of staffing and supplying the areas. Some parks have been imposed with a heavy hand, sparking local resistance and providing lessons for the protected areas the world still needs to create. An innovative array of park categories has emerged during the 2000s, particularly in the Amazon, to accommodate the inevitable interaction of people and nature as both population and the protected acreage grow.

At least half of intact forest landscapes, around 1.4 billion acres (570 million hectares), fall outside both protected and Indigenous areas. The megaforests need more protected areas with more funding and training for staff. The cost for greater protection of the forests is a bargain: $1 to $2 per acre per year.

To make megaforest conservation work—whether Indigenous lands, parks, or other areas—the single most important factor is limiting roads. In the tropics, almost all deforestation takes place along roads or big navigable rivers. Even where outright deforestation is minimal, as in the Congo, roads open the jungle to hunters, and defaunation ensues. Boreal roads are vectors for overhunting and fire and can block water from moving through wooded wetlands. Because remoteness and size hamstring megaforest policing, the more roadless they are, the less lawless.

The pursuit of roadlessness in the United States goes back to the 1920s when American ecologist Aldo Leopold, then a Forest Service official in the Southwest, realized that roads and intact ecosystems didn't mix. The heavily roaded lands under his care were ailing, and trout streams were drying up. He scoured the map for a place still free of roads and in 1924 set up the first national forest wilderness, called the Gila, in a mountainous corner of New Mexico. The tradition reached its apogee in 2001, when the Forest Service protected all the nation's remaining 58 million roadless acres of national forest, including America's last handful of temperate intact forest landscapes, in southeast Alaska.

These strategies hold promise for a future in which vibrant, planet-sustaining forests will benefit the lives of descendants so distant they won't know our names. But saving the megaforests requires more than strategy and tactics. This is a job that needs to be done with feeling. In the modern world, the beings and features of the forests, and even the forests themselves, are commonly thought of as objects. We humans are separate subjects acting upon them, and nearly all verbs, if we think of this grammatically, are ethically tolerable. Cut, dig, clear, gather, manage, thin, burn.

This separation is rare among forest peoples. Little Anastasia Momo's family may go back 50,000 years in New Guinea's forests. Like most clans here, the Momos have an origin story reaching into a prehuman realm. They identify black cockatoos, like the ones flying around us on this hillside in the Tambrauw Mountains, as ancestors. Nearby clans ascribe that role to tree kangaroos or masked snakes, creatures with whom long unbroken lines of ancestors have shared the forest shade. People we have met over the years, and others we interviewed for this book, repeat refrains about kinship and obligation to the various forms of nonhuman forest life that sustain them. Accounts abound about times when people and animals were fully conversant in the same language. Some still talk. For modern humanity to keep the megaforests, and with them the one planet we know of that has any forests, we need to care for the world as if it is family. We need to attempt a grammar in which subject and object, people and everything else, are the same. In a material and evolutionary sense, of course, we absolutely are.

1

The Forest System

Shortly before dawn on October 9, 2017, John looked out his window and saw a red glow in the east. Minutes passed and, oddly, as if the sunrise had stalled, the glow didn't brighten. He walked out in the yard and held out a hand. Gray flakes of various sizes fell soundlessly. He fished a singed magazine page out of a rhododendron. A skier from another decade, looking delighted, smiled up from the page, inviting the reader to vacation in Bend, Oregon.

The sky lightened to a miasmic yellow. Most of the streets of Sebastopol, California, were empty. The high school, however, was buzzing with armored vehicles, National Guard troops, and a flatbed truck from which neighbors were unloading crates of water, tangerines, and energy bars for the evacuees starting to fill up cots in the gym. Santa Rosa, a small city 6 miles away, was on fire. K-Mart and Trader Joe's and whole neighborhoods were going up in flames. Over the next few days, as the fires raged, everyone learned what an N-95 face mask was, hosted evacuated friends, signed up for emergency alerts on their phones, helped out at the shelters, and watched the sheriff's daily briefings on the Internet.

The fire that started on Tubbs Lane in Calistoga was like nothing the region had ever seen. It had crackled and fumed its way swiftly through forests of oak, fir, bay laurel, and buckeye, over the hills in the night. It raced downhill, which is hard for fires to do, hurling fireballs to the south and west, and eventually laid waste to block after block of Santa Rosa. Most of Sonoma County's 500,000 residents came away physically unscathed, but with a permanently altered awareness of the world.

We are living climate change, fully immersed in the future that we were only talking about until recently. In 2020, the routine continued for the fourth straight year, fires handily setting a new California record for charred acreage, alternately coloring the world orange, sepia, and, as one Bay Area writer for the *New York Times* put it, "yellow-gray, like a smoker's teeth." One September Wednesday San Francisco went red, like a city under a darkroom light. Equivocation is gone from the media accounts and scientific discussions; the drought-baked landscape and fires that rip through it are results of a changing planetary reality.

Hopefully it is enough to provoke changes in the way we're living. Societies around the world need to adapt, each according to its own particular circumstances, but generally transforming energy systems, transportation, manufacturing, and what we eat. The human population needs to stabilize and start to decline in order to cut the amount of energy, food, transportation, and other things we ask our finite planet to provide. There are glimmers of progress on all of these fronts.

To meet the climate challenge, there's one other essential task we have to accomplish: save the world's biggest forests. The planet is a linked physical-biological system in which large wooded expanses keep both local and global conditions stable and livable. They metabolize the carbon our economies so relentlessly put in the air in a process that circulates life-giving water around our landscapes. This physical work is accomplished with a biological mechanism involving trillions of organisms belonging to millions of distinct species in a constant whir of transacted matter and energy, moving from one being to another, from earth to sky and back.

Our world keeps its carbon in four places. One is the lithosphere, a term that comes from Greek words meaning "rock ball." Carbon made solid by ancient photosynthesis is stored in Earth's rock layer in combustible forms like oil, gas, and coal, as well as other substances, such as graphite and diamonds. The second place is the atmosphere (from the Greek, meaning "vapor ball"), where the element mostly takes the form of carbon dioxide gas. The third place is the hydrosphere, the planet's surface water, 97 percent of which is ocean. When seas absorb carbon dioxide from air, their water becomes more carbonated, like an oh-so-slightly fizzier soda ($H_2O + CO_2 = H_2CO_3$, carbonic acid).

Finally, there is the biosphere, the layer of living stuff between rock and air. Plants slurp carbon dioxide molecules through tiny pores, cleave off the carbon, and build themselves out of it. Carbon makes up around half of plants' mass. Growing things drop leaves, cones, seeds, flowers, branches, and eventually trunks and stems onto the ground. Some decomposing biomass goes back into the air and some into the soil, the proportion depending on the speed of decay. Of the carbon that is buried, some is compressed over the eons into the fossil fuels we're now quickly burning. As for the carbon that stays topside, vegetarian animals eat the plants and incorporate the carbon into their bodies and are, in turn, eaten by carnivores, the apex carbon collectors.

The distribution of carbon in these four realms has varied over time. During periods of rapid cooling, Earth made a lot of plants into coal. When the planet had extensive shallow seas, the ocean floors became vast graveyards for tiny plants and animals that were eventually transformed into oil and gas. Over the 200,000 years of our species' existence, the atmosphere's CO_2 has oscillated between 170 and 280 parts per million. The last 10,000 years are a period called the Holocene Optimum, a time when temperatures have been very stable, a hair below their current levels. This has been the climatic stage on which the human dramas of agriculture, industry, and explosive population growth have played out. These plot twists were supported by the Holocene Optimum and are now ending it as we withdraw carbon from the biosphere and lithosphere and deposit it into the atmosphere and waters.

The official scientific body that reports on this adventure into the climatic unknown is the United Nations Intergovernmental Panel on Climate Change, or IPCC for short. Thousands of scientists contribute to its bulletins, which address the question of what our planet will be like in the future. That depends on how much more carbon we transfer from the ground to the atmosphere. A pair of reports, released in 2018 and 2019, say we need to save forests to save the planet. One report concludes that the world will be in dramatically better shape if we limit long-run warming to 1.5°C (2.7°F) rather than 2°C (3.6°F), and the other explains how our treatment of the planet's land needs to improve in order to accomplish that. Tropical forest loss alone emitted around 5 billion metric tons of CO_2 annually in the first decade of the 2000s. For a sense of scale, that's more than all emissions from the European Union over the same period. This jet of greenhouse gas from the biosphere would be even more troubling but for the fact that the *uncut* tropical forests reabsorbed around half of it.

The IPCC advised that forest loss needs to stop completely by 2030. Thereafter, the panel prescribes increasing the wooded area by up to 85 million acres per year until 2050. That would help us stay within an atmospheric carbon budget for 1.5°C warming, with less risk of fraying ecosystems and major social upheaval and less need to resort to high-risk technologies such as nuclear energy. A separate analysis found that stopping tropical deforestation would reduce global emissions by 16 to 19 percent.

For a couple of decades, climate talks were organized around a warming ceiling target of 2°C. This round number emerged from an IPCC led by physical scientists, unaware of the full scope of the violence 2 degrees of warming will do to the planet's biology. We know from past climate change that species respond idiosyncratically, each at its own pace and in its own direction. They don't move as entire biological communities. That pulls ecosystems apart, and species either perish or regroup into new sorts of nature. For example, the partnership between the coral animals and algae that creates reefs is unraveling; warmer water causes the coral to expel the algae on whose photosynthesis its life depends. Reefs collapse and fish are left to improvise or die.

On land, the coexistence between native bark beetles and their conifer hosts is faltering. Warmer winters let insects survive in unusual numbers, and drought limits trees' production of sap toxins with which they fought off the insects in wetter times. There have been massive tree kills in the American West, British Columbia, Europe, and Siberia. One researcher in Utah reported beetles, having devoured all the pines in the area, tucking into telephone poles. The drama of the damage we're already seeing at one degree above preindustrial levels throws into relief what the forests might be in for anon. The IPCC says that going from 1.5°C to 2°C will take terrestrial ecosystems from "moderate" impacts to "high" ones.

The 2-degree goal also promised a perilous future for small island nations, forty-one of which make up a negotiating bloc at the UN climate meetings. The situation of the Marshall Islands, Kiribati, the Maldives, and others is especially dire because there is nowhere for people to flee. But in terms of sheer numbers of people whose homes will become uninhabitable, sea-level rise promises to wreak the greatest devastation on six Asian mainland countries. At the urging of the island leaders, the Paris gathering of the climate pact's parties in 2015 dialed the temperature goal down to 1.5°C.

All forests can help. But large forests are of supreme importance for the climate. Intact forests are 20 percent of the tropical total and store 40 percent of the aboveground forest carbon in the low latitudes. New research led by Sean Maxwell, of the University of Queensland, and eleven collaborators suggests that the carbon benefit of intact tropical forests is six times greater than the IPCC and others have estimated to date. That's because in the years after a big forest is broken up by roads or farms, its edges dry out and winds whistle through, blowing over big trees. Fires invade it more readily, and overhunting eliminates animals that disperse seeds. And on top of all the carbon vaporized from the space actually deforested, over the next several decades the climate will be stuck with 14 metric tons of extra carbon per acre that the lost tropical forests would have absorbed had they remained standing.

The consequences of fragmentation are similar in the boreal forest. Even small amounts of deforestation create hot, dry forest edges and warm the forest interior, far from the bits actually cleared. That makes the understory highly flammable. Michael Coe, climate scientist at the Woods Hole Research Center, is an Amazon expert who collaborated with temperate and boreal forest specialists on a 2020 study of forest-climate dynamics across all latitudes. He says that fragmenting the boreal can lead even more directly to the incineration of the remaining trees than is the case in the tropics. "Any kind of an edge, it doesn't have to be a big edge, causes a problem," says Coe of the boreal forest.

When forests are kept intact, they deliver a double climate benefit. They cool the planet, thanks to CO_2 removed from the atmosphere, and cool the local environment through the processes of evaporation and transpiration. Evaporation is the familiar process of liquid water, on all the forest surfaces in our case, warming and turning into vapor. Transpiration is the exhaling of vapor that originates inside the leaves and escapes through pores. The combined process is called evapotranspiration. Just like sweating cools people, as water turns to vapor it absorbs energy and cools the surrounding environment. You can feel this air-conditioning in the forest interior, which is cooler than a treeless shady spot, say, under an awning.

Tropical and boreal forests have different rhythms for harvesting and storing carbon. The tropical forest grows riotously all year, minting solid biomass from CO_2 and shaping it into trees, shrubs, ferns, ground covers, orchids, and other plants. Its pollinators, seed dispersers, and bacterial and fungal partners are of unfathomed number and diversity. Fallen leaves and wood decay into a thin layer of soil whose nutrients rainforest roots tap immediately to grow more plant matter. Liquid water is available year-round to support plant growth, evaporating and transpiring continually from plants into clouds that coalesce, move, grow heavy, and spill onto another patch of woods that does the same thing all over again.

The boreal forest, by contrast, is a patient, seasonally photosynthesizing interface between the sky and underground carbon caches. In the northern parts of the boreal, trees can take many years to get as tall as a person. Throughout the ecosystem, they grow during a short summer and continually shower the forest floor with needles, leaves, cones, and twigs. Some material falls into oxygen-deprived waters and changes extremely gradually, like specimens preserved in laboratory jars. In the winter it's too cold for microbes to process the vegetation into soil. Vegetative "sediments"

Amazon forests, like this one in the Yanomami Indigenous Territory, create their own rain and amass carbon. *(©Sebastião Salgado)*

are thus packed into ever thicker deposits of soil and the proto-coal called peat, a semi-decomposed layer that comprises 47 to 83 percent of carbon in boreal ecosystems.

On average, 95 percent of boreal plant carbon is underground, compared with 50 percent for tropical forests. Average belowground carbon in boreal forests is somewhere between 154 and 197 metric tons per acre, up to five times the average in tropical forest soils. A 2015 study found that boreal carbon stocks were four times as extensive as the IPCC had estimated in 2007 and twice as large as calculated in a more thorough 2011 inventory. How could so much carbon have been hiding? Researchers weren't digging deep enough; soil carbon estimates had been based on the first 3 feet (1 meter) underground, and there is lots of carbon buried deeper. The boreal average is somewhere between 4.25 and 7.5 feet deep.

Intact forests are only now being fully recognized as central to the climate crisis and its solutions. In 1992 the UN Earth Summit in Rio de Janeiro produced a climate treaty, which largely excluded forests, and one for biodiversity, which embraced them. Among those fighting to keep forests out of the climate accord were some environmental advocates, who argued that forest carbon was hard to measure and that giving countries credit for dodgy forest emission reductions might permit very real increases in industrial CO_2 pollution. The measurement problem was largely solved by 2010 thanks to advances in aerial and satellite technology, wide availability of data, and improved computing power. At the same time, tropical forest countries started playing more prominent roles in treaty negotiations. In the last five years, as the urgency of the climate crisis has heightened, researchers have begun to confirm the surpassing climate advantages of very large forests.

Since long before the climate crisis gained wide attention, biologists have been fascinated by intact forests because they are healthy, complex systems with countless species, not because all that nature is made of carbon. They have predation, pollination, seed spreading, and procreation all happening naturally and in profusion. They have troops, colonies, packs, and pecking orders; microfauna, megafauna, intrepid migrants, and entrenched residents. Harpy eagles eat spider monkeys, grizzlies eat salmon, tree snakes eat tree frogs, pitcher plants eat ants, and ants farm fungus. As we hinted earlier, a biologist in a large tropical forest can reasonably hope to find a living thing no scientist has previously encountered. A three-year study concluded in 2018 in the Madidi National Park, in Bolivia, found 124 new species, including 84 previously unknown plants, 19 fish, 8 amphibians, 5 butterfly species (plus 8 subspecies), along with 4 new members of both the mammal and reptile classes.

Biologists have long known that larger ecosystems have more species. Charles Darwin noticed this relationship while visiting small islands, such as those in the Galápagos archipelago, where he famously found some quite peculiar species, like the marine iguana, but somewhat less famously noted the rather small number of species overall. In a small, groundbreaking 1967 book, *The Theory of Island Biogeography,* Robert MacArthur and E. O. Wilson proposed an elegant equation that described the relationship between the size of islands and the richness of their biota. The young Drs. MacArthur and Wilson discussed actual islands and also used them as a metaphor for the fragments of habitat isolated in a landscape altered by humans.

Fragmentation hadn't previously aroused much scientific interest or environmental concern, because fragments lost their species gradually. The island comparison brought the issue into focus. A spirited argument broke out in the 1970s over whether more biodiversity could be protected in one large area or in a few smaller ones with the same total extent, strategically positioned to capture variations in habitat. This became known as the "single large or several small" (SLOSS) debate.

The SLOSS tempest raged among scientists who lacked data to corroborate their respective stances. The only information was from a single site, a tropical forest island called Barro Colorado that was created in the artificial Gatun Lake when the Panama Canal was completed in 1914. The site in the formerly US-administered Canal Zone is still managed by the Smithsonian Institution, and for a rainforest field site, it has exceptional infrastructure and accommodations. Barro Colorado's managers accurately declare it "the most intensively studied tropical forest in the world." By the 1970s, it was clear that species were declining on the 3,853-acre island. The dynamics of a single island, however, provided a tenuous basis on which to predict how isolated forests would fare in the seas of pasture and crops spreading across the world's continental forest zones.

In 1973, Tom took up a post as head of programs for the US office of the World Wildlife Fund. He realized that WWF needed to know more about habitat fragmentation. How else could they determine whether their conservation projects were big enough to save species? Then he remembered that Brazil's forest law required landowners to leave 50 percent of Amazon rainforest standing as they mowed down the rest for cattle ranching or crops. He proposed to the United States National Science Foundation (NSF) in 1976 that a Brazilian landowner might be persuaded to leave that

50 percent in a configuration that would provide a giant forest fragmentation experiment. With the NSF's backing and that of the Brazilian National Institute for Amazonian Research (INPA), in Manaus, he approached the Brazilian bureaucracy in charge of fomenting cattle ranching with a request: ask ranchers to leave their required reserves in squares of various sizes surrounded by pasture. The agency agreed.

This experiment began in 1979. It ended up with five plots measuring 1 hectare, four at 10 hectares, and two covering 100 hectares, which convert to 2.5, 25, and 250 in acres. Matching control plots in continuous forest were also established. By 2002, the project had produced a simple answer about fragmentation: large intact areas are very important, the larger the better. Even the 250-acre reserves were too small for forest interior bird species, half of which vacated these patches in less than fifteen years. The edges were hotter and drier, with great mats of desiccated leaves from trees either dying or losing foliage to wind. There were more vines, thicker undergrowth, and fewer mushrooms.

Species that need continuous tree cover decamped. Black spider monkeys, for example, who move fast through large areas of forest eating fruit from widely spaced trees, abandoned all the forest fragments immediately. They stayed in nearby continuous forest. Howler monkeys, by contrast, are leaf eaters and not particularly choosy. They remained in all the fragments. The white-plumed antbird, so named for the spiky crest between its eyes, could not persist in the fragments. Antbirds follow raiding ant armies and eat the bugs flushed out by the lethal column. While 250 acres is sufficient territory for one ant colony, each colony marches only about a week per month. So, to avoid going hungry for weeks at a time, the white-plumed antbirds need to follow several colonies on a rotating basis. The 250-acre fragments were at least three times too small for the birds. No antbirds means no antbird droppings, which deprives shimmering blue-and-black skipper butterflies their sustenance. They left, too.

Birds such as the black-tailed leaftosser, which finds insects by turning over leaves on the forest floor, also ran into problems. The forest fragments were pummeled by wind, which felled trees up to a quarter mile from the edge. Resulting gaps were filled by trees in the *Cecropia* genus,

Black spider monkey in intact Bolivian rainforest.

which you can see along almost any Amazonian roadside, riverbank, or regrowing pasture. The *Cecropia* leaves are like lobed umbrellas that can easily measure a foot across, too large for the leaftosser to upend. Most insectivorous birds and bats, along with arboreal mammals, dung beetles, wild pigs called peccaries, and orchid bees found even a narrow clearing insuperable. A couple hundred feet of treeless ground, typical of a highway, was enough to prevent their using the forest fragment habitats.

At least four frog species that live in the wallows created by white-lipped peccaries vanished from the fragments; the pigs that dig their pools wouldn't use the forest islands. These amphibians were replaced by "generalist" frogs common in cattle pastures.

The forest fragments project, with its emblematic squares, spawned a field of study focused on what happens when big forests are made smaller. Its findings firmly established forest fragmentation as an urgent environmental problem. Hundreds of Brazilian and other graduate students have earned advanced degrees studying plants, animals, soil, and carbon in the original plots. Many more have investigated the unscripted fragmentation of forests across the world. This body of science corroborates Darwin's original observation: intact nature has more diversity than nature in pieces.

So, megaforests are both more biologically diverse and more heavily stocked with carbon than fragmented ones. That raises a tantalizing question: is there a universal mechanism that links diversity and carbon accumulation?

In intact Congo forests there's an explicit connection: elephants. They snack on small trees, which allows bigger ones to occupy more space, catch more light, and grow denser wood than the browsed stems would have done. Elephants love big fruit, which comes from big trees. When you come upon their loamy dung in an African forest, you'll often find it sprouting future forest giants. Elephants are thriving in Nouabalé-Ndoki National Park, in the Republic of the Congo. According to a 2016–2017 census, the park had over 3,000, or one for every 3 square miles of jungle. In a less scientific measure, fleeing forest elephants is a routine part of visiting the area; during a visit in October 2019, John twice ran for his life. This intact forest has 15 percent more carbon than a comparable patch of Congo woods where elephants have been eliminated. The finding is remarkable and, in a way, surprising, given the roomy feel of a forest sculpted by elephants.

Another study in the vicinity of the Nouabalé-Ndoki National Park showed that hunting in general—not just for elephants—reduces the carbon held in plants. A forest free of logging and hunting averaged 455 metric tons of biomass for every 2.5 acres (1 hectare). Logged but unhunted forest had 358 metric tons, more than 20 percent less. The third category was both logged and hunted. These areas had largely lost their elephants, leopards, gorillas, wild pigs, and forest antelopes, and had 301 tons of biomass and a remnant fauna consisting of "squirrels and small birds." Most of the missing animals were herbivores. Apparently, the more plant life they eat, the more there is.

Across the Atlantic, Amazon orchid bees live in undisturbed forests. These flying gems come in a range of metallic greens, blues, purples, and oranges. Some look like they're straight out of a superhero comic, with shiny segments, including heavy "boots" on males' hindmost legs, and tongues longer than their bodies. The males alight and quickly squeegee orchid scent off the flowers with wide brushes on their forelegs. They stash the cologne in packets in those hind-leg boots for wooing burly females. The female orchid bees pollinate one of the forest's biggest trees, the Brazil nut. It has yellow-and-white flowers that last one day only and have heavy lids. Orchid bees depend on their developed musculature to prise open and pollinate the blossoms. The resulting fruits are packed with succulent, oily Brazil nuts encased in a woody shell that's nearly impenetrable.

Unless you're an agouti. The large rodent, routinely hunted across the basin, employs a viselike jaw and chisel teeth to get at the Brazil nuts, but only eats a few from each pod, hides the rest, and forgets where most of them are. The forgotten seeds grow into new Brazil nut trees. If people disturb intact bee habitat or overhunt agoutis, the dense and voluminous Brazil nut trees will vanish from the landscape, leading to a carbon loss.

Overhunting of seed-dispersing animals, such as spider monkeys, woolly monkeys, and tapirs, could cost the Amazon 2.5 to 5.8 percent of its stored carbon. In some regions, declines of 26 to 38 percent are possible. One study that involved heroics in data collection showed the importance of fauna for both forest carbon maintenance and restocking. Tapirs are the largest land animals of the South American rainforest. They are tasty and funny looking and shy. And they are a serious actor in the megaforest's capacity to sock away carbon. The experts estimated that tapirs defecate 1,194 seeds per acre per year in undisturbed forests and up to three times that many in forests recovering from human disturbances.

Links between forest biodiversity and carbon storage abound. Further, it is clear that a full complement of species makes forests resilient to climate change. That's because if a species should succumb to the shock of changed conditions in a forest that still has its full array of native biodiversity, it is more likely to be survived by some other plant or creature, as the case may be, that can fill the ecological gap.

So far no one has identified a single rule of nature that links biological diversity to forest carbon maximization. There are instances where the two go together, and there are vast forest regions in which both are remarkably high. But nature needs insects, birds, and bats that pollinate and spread the seeds of less dense trees. And there are creatures that thrive in a forest with sunny openings in the canopy; gorillas seek these spots to feast on fast-growing soft-stemmed plants.

The urgency of climate change, however, has compelled many scientists, economists, and environmentalists to think about how much carbon there is in everything, even in the bodies of elephants. Carbon, the element, becomes a currency, the unit of measurement in a chemical accounting system we use to chart survival paths for civilization. The peril, of course, is that this carbon myopia conceptually distills the intricacy of a forest ecosystem into a colorless idea small enough to fit in a beaker.

Some studies show that when animals are gone, the forest sheds plant carbon. But what do we make of the forest animals whose removal has a negligible impact on carbon? Do we write off the gibbons of certain forests of Southeast Asia, the pollen and seeds of which are wind-borne? What will become of boreal creatures, even famous ones like caribou, if they are found to be contributing too little to the production of peat?

Aspens thrive in the boreal forest of northern British Columbia, near the Yukon Border.

An engineering mindset may also lead us to muse whether the forest might be force-fed a bit more biomass. Some scientists say it might indeed, with performance-enhancing genes that augment photosynthesis and carbon transfer from plant to soil; it's possible to juice the jungle. Another idea, about which scientific papers are written, is to cut down the boreal forest—all of it—so the snow reflects sunlight in winter. This wouldn't work because it's impossible to cut down the whole boreal and keep it from growing back, and razing it incrementally would emit more carbon than the reflective cooling could make up for. In any case, says Michael Coe, of the Woods Hole Research Center, "An engineering solution that destroys biodiversity is a bad idea. There are always unintended consequences."

Carbon myopia clouds the true meaning of the orchid bee and the Brazil nut. Big forests are a linchpin in a planetary system. They are vivid stages for stories about energy and matter that we describe severally with our physical, biological, and chemical sciences, but are really a single story whose intricacies and meaning we don't fully understand. Orchid bees make Brazil nuts, breed agoutis, take carbon from the air, breathe water back into it, make clouds that make rain a hundred miles away that feeds a stream, where a catfish, having migrated from the mouth of the Amazon, is caught by an otter or by a person, surrendering its protein to enliven the woods. The bee makes all these things, and these things make the bee. Losing the forest would change more than the reading on the thermometer. Wind, rain, fire, and ocean currents would be rewritten. If we lose too many trees, everything changes. The weave unravels.

Wild megaforests offer a simple, *existing* solution to achieve carbon storage, local climate stability, and the survival of various, individually useful, mysterious, and beautiful carbon-based organisms. The woods require no engineering and are free of surprise side effects. Thrift, common sense, risk aversion, and an appreciation for natural beauty all argue for this solution. A smart approach to managing the planet, which we must, is to leave some parts, including the megaforests, mostly to manage themselves.

2

Mapping the Root Forests

You would think something as large as a megaforest would be easy to spot. In broad strokes, that's true; airline pilots, astronauts, and readers of satellite images could tell you that there are big tropical forests in the Amazon, Central Africa, and insular and mainland Southeast Asia. And they could attest to the mammoth forests in the north of Russia and North America. The eastern United States, sections of southern China, and Scandinavia also look pretty well treed. Outsized blotches of green, however, don't disclose which forests are laced with logging roads or peppered with buildings and mines, and which ones, on the other hand, are unbroken refuges for creatures that demand a lot of space or deep forest interiors to thrive. A cursory look doesn't reveal which big green places, by virtue of their physical integrity, have large and stable carbon stocks, healthy microclimates, and vital rivers.

Until the late 1990s, people talking about big, exceptional forests used terms like "primary," "pristine," and "virgin." These expressions carry some misleading connotations of immutable prehuman purity; people have been dwelling in the world's forests for tens of thousands of years, and they are still there. What's more, the lack of a clear definition made primary, pristine, and virgin forests unmappable. In 1997, the World

Resources Institute (WRI) in Washington, DC, took a stab at a definition and a map. With a touch of Wild West romanticism, they identified the world's "frontier forests." The criteria were nuanced and thoughtful and appropriately focused on ecological integrity. They included forested ecosystems big enough to support their widest ranging species, places mainly shaped by nonhuman influences, with nearly all native trees and nearly a full complement of plants and animals that would naturally occur. A rather limited set of forests, clustered in the tropics and boreal zones, qualified.

Around the same time, events along the Russian-Finnish border were creating an urgent need for a good map of big and exceptional forests. Russia and Finland have long sparred over their boundary. The most recent conflict was the 1939–1940 Winter War, which resulted in hundreds of thousands of casualties. The Soviets prevailed and established an unofficial 25-mile-wide demilitarized zone along the frontier, off-limits to logging and other development. Over five decades the border area recovered and matured into a splendid forest known as the Green Belt of Fennoscandia. In a 1984 episode of BBC's *Living Planet*, a nimble David Attenborough puts on a helmet and goggles, climbs a very tall ladder, and pulls a great gray owl chick from its nest in a hollow Finnish pine. The tree is within view of the Russian border, beyond which, Attenborough explains, there is high-quality habitat to which the owls return in times of scarce food.

A few years after the naturalist's ladder caper, the Soviet Union collapsed, and these plum timberlands were opened up for business. European loggers started buying Russian trees to supply businesses, including the Swedish assemble-it-yourself furniture maker Ikea. At first no one noticed. "What was out of sight was out of mind, and much of what was happening was out of sight," said Lars Laestadius, a Swede who worked with WRI at the time as a forest expert. Eventually, European and Scandinavian environmentalists cottoned onto the destruction, reacting with publicity campaigns that targeted Ikea, among others.

Greenpeace activists and Ikea representatives started talking. The company was not about to give up making furniture out of wood, and the newly opened Russia, right next door, had the world's biggest forest. The parties agreed that wood shouldn't be sourced from certain special forests. Ikea called them "natural," while Greenpeace used the term "ancient." There was hope, however, that they were talking about the same places. Ikea asked for a map and, with Greenpeace's blessing, provided a grant to WRI, creator of the frontier forests map. "I was given a bag of money and told to produce operational maps for Russia," Laestadius says. It quickly became apparent that the frontier forests maps weren't "operational" for sourcing wood. "They were too coarse and had no local roots . . . As an old Soviet Studies major—before going into forestry—I wanted the maps to be made in Russia by Russians. Otherwise, they would have no impact." Laestadius also knew that the American baggage carried by the "frontier" label would undermine the embrace of these forests in Russia.

A small group of Russian environmental activists went to work on new maps in 1998. Alexey Yaroshenko, a young botanist working for Greenpeace in Moscow, joined Peter Potapov, an ecologist at the Biodiversity Conservation Center, a Russian nonprofit, and Svetlana Turubanova, an ecologist-volunteer at Greenpeace (who would later wed Potapov). They were advised by Olga Smirnova, their former professor and an oracle on old-growth forest ecology at the Russian Academy of Sciences.

To answer Ikea's question, the group came up with the first-ever method to identify which forests were least affected by industry, infrastructure, farms, and settlements. These were the places where activists wanted to ban logging. The big-picture thinker behind the strategy was Yaroshenko. Now in his mid-50s, he looks younger and has an easy confidence. Speaking from his dacha by video call in mid-2020, he said they were not looking for pristine forests that had eluded all historical interactions with people. "Intact forests are not completely wild nature that never knew the influence of people. No, that's not the idea. They are the last remnants of forest that were in equilibrium with an old type of human influence. They avoided modern industrial development." Yaroshenko had read a Sierra Club magazine article in the late 1990s that used paper maps to pinpoint the wildest nature on the planet. Those were exactly the sorts of places he wanted to tell Ikea and their suppliers to stay out of. Accordingly, his group's first step was to unfold old-school maps and digitize places that already had roads, railway lines, towns, and other industrial footprints.

The next step was to examine publicly available satellite images. At the turn of the millennium, the only satellite providing public images of Russia's forests was an orbiter called *Resurs*, meaning "resource." Each pixel represented a square of ground around 500 feet (150 meters) on each side, precise enough to identify large clear-cuts and farms but inadequate for smaller disturbances, like roads. So the team selectively bought as many images as they could afford from the Landsat satellites operated by NASA and the US Geological Survey—which were twenty-five times more precise than those from Resurs—and went to sixty-seven spots on the ground to see how real conditions squared with the pictures taken from space. They mapped all tracts that were at least 50,000 hectares (around 125,000 acres) and at least 10 kilometers (6 miles) wide; wormlike strips didn't qualify. These big forests were designated intact forest landscapes, or IFLs. The landscapes could include wetlands, alpine meadows, lakes, mountaintops, and other naturally treeless parts of healthy forest ecosystems.

We met Peter Potapov and Svetlana Turubanova at their lab at the University of Maryland in College Park in 2019. Dark-eyed Turubanova is Komi, a people native to the lands on the western slopes of the Ural Mountains, the north–south cordillera that marks the boundary between Europe and Siberia. Her doctorate in paleoecology focused on old-growth forests. "I'm from a very forested area and I love to go mushroom hunting!" Turubanova laughed. "For me it's natural, it's a forest, it's my home, I want to protect it." Sandy-haired Potapov has a PhD in ecology and natural resource management and deep technical expertise in geographic information systems. He doesn't hesitate to use nontechnical terms like "evil" and well-chosen expletives to describe the companies destroying the forests. They both share with Yaroshenko a sturdy twist of realism and ideals. They see the world as it is with cartographic precision and envision how it should be through the lens of their personal environmentalist values.

They credit their mentor, Olga Smirnova, for this mindset. She took them on field trips to the lost world of Russia's "root forests." This is the

Svetlana Turubanova, Peter Potapov, and colleagues created the first global map of intact forest landscapes.

term they offer when asked for the Russian expression for extraordinarily intact, natural woods. The students were entranced by the complex webs of organisms in the remote, unlogged woods Smirnova chose. She taught them that the natural forest, in Potapov's words, "is a living thing," not just some trees projecting from the ground. Smirnova is 80 now and has been working in Russia's forests for over sixty years. As we discussed forests on a video call in 2020, she cited academic papers from the 1920s and 1930s. The smiling Dr. Smirnova looked like a cute Russian grandma in her apartment. But later she sent fieldwork photos in which the octogenarian was in head-to-toe camo rain gear, teaching a group of Russia's youth in the hail and snow of early summer in the central Ural forests.

Smirnova explained that root forests have eight earthworm species to the two found in logged woods. And she rhapsodized about a certain hot-pink flower—casting about for the English name—that fill old-growth forests at around the time of our June conversation. Peonies. In an email before our video call she emphasized the complexity of an old, intact forest. "This is a complex combination of glades with grasses, shoots, and young trees; with trunks of dead trees filled with life. There are coexisting soil animals, mushrooms, mosses, microorganisms, and other creatures." Smirnova taught Potapov and Turubanova that some forests were more fully vital than others. They were likely to be the big ones.

The Russians chose the 125,000-acre threshold for IFLs because it's a size that can accommodate natural processes, including small fires, trees felled by wind, and protection of lakeshores and riverbanks. Forests this size can play a role in the life cycles of species that need a lot of space, such as tigers, reindeer, bears, and wolves. And the forest interior is well protected from disturbances. Their second rationale was that big areas tend to be less expensive to conserve; one large forest is easier to patrol than many small ones. Asked if it might make sense to set a smaller size threshold in other forest types, Potapov and Turubanova provided a third reason for the 125,000-acre lower limit: mapping smaller intact areas on continental landmasses like Russia's proved completely impractical.

Teaming up with several Russian organizations and WRI's Global Forest Watch program, the original group expanded their map to include all of European Russia in 2001. The next year they mapped the IFLs across Siberia and the Russian Far East, covering the entire Russian megaforest.

The potential of this analysis to tell a story about the whole planet's forests wasn't lost on geographer Matt Hansen, who runs the Global Land Analysis and Discovery (GLAD) lab at the University of Maryland. He hired Potapov in 2006, shortly before the US government gave the public free access to its three-decade archive of Landsat satellite images for the first time. NASA had tried to make money from the images in the 1980s, but at the exorbitant price of up to $4,400 for one image, buyers were few.

NASA cut the price to $475 in 1999, and then started to offer the images free in late 2008. Downloads exploded. Before this open-access policy, a global forest map would have cost at least $2 million, just in raw data. With free images and computing power on loan from Google, a ten-year-old firm at the time, Hansen's team could readily assemble the satellite pictures into a single portrait of the inhabited continents, a mosaic of 150 billion pixels, each representing a 98-by-98-foot (30-by-30-meter) square.

In 2008, Potapov, Yaroshenko, Turubanova, and twelve coauthors published the first peer-reviewed map of world's intact forest landscapes. It showed an area of 5 million square miles (13.1 million square kilometers) of IFLs in 2000, 84 percent situated in the tropical and boreal regions. The intact forest landscapes covered 23.5 percent of the planet's "forest zone" (defined as places where there were trees growing in the year 2000) and 2.6 percent of Earth's land area. In a high-profile *Science Advances* paper in 2017, Potapov, Hansen, Turubanova, and others updated the map. They found that around 10 percent of IFLs had been lost or fragmentated from 2000 to 2016, due to logging, fire, expanding farms and ranches, and the mining, oil, and gas industries.

Not everyone loves the IFL concept. The main objection is the bar it sets for size. Scientists point out that the 125,000-acre standard filters out smaller areas of extreme importance for biodiversity conservation. The highly fragmented Atlantic coastal rainforest of Brazil, for example, is packed with plants and animals not found elsewhere on the planet, known as endemics. The once massive rainforest has a total of three modest IFLs out of the globe's 2,000. The Western Ghats of India, another center of forest biodiversity, has none.

On the other hand, the IFL threshold is in some places too small. Many wide-ranging and migratory species and ecological processes need far more than 125,000 acres. Canadian scientists note the immense territories nibbled at by woodland caribou and the vast scale at which landscape-shaping dynamics, such as fires and species migrations, take place. Conservation planning, including identifying which forests to protect, has to be done on a bigger scale than IFLs to embrace these species and events.

Like all good proxies, size is handy but imperfect. MacArthur and Wilson's *Theory of Island Biogeography* made clear that larger island size doesn't *cause* there to be more species. Size is associated with certain ecological properties that lead to diversity, including habitat variety, population sizes, and migratory opportunities. One response to the size dilemma is to do away with the minimum threshold but screen for forest age and quality. In 2018, Turubanova led a group that identified forests they called "primary" in three countries that account for most of the world's tropical jungles: Brazil, the Democratic Republic of the Congo (DRC), and Indonesia. By "primary" she doesn't mean original or untouched. Primary forest includes forest that has remained unchanged over the several decades of available *Landsat* imagery and *looks*, in the most recent images, like unmolested tropical forest should. Its canopy texture is bumpy because trees are of various heights, including exceptionally tall "emergents," and generally darker in color compared with forest substantially impacted by people. These traits vary across the continents, so the analysis is customized for each zone. The Amazon has more palms, for instance, and the forests of Borneo naturally grow taller than any others in the tropics. Turubanova's team also looked at a sample of ultrahigh-resolution images to make sure that what looked primary in the *Landsat* frames was primary in real life.

The resulting map located a lot of primary tropical forest that was excluded by the original IFL map. In Brazil, 30 percent of Turubanova's primary forest had not been included among the intact forest landscapes. In the DRC, 40 percent fell outside the IFLs. And in Indonesia, a whopping 66 percent was excluded from the maps that applied the 125,000-acre minimum. Put another way: there is about three times as much relatively wild "root" forest as there is intact forest landscape in Indonesia, a populous archipelago-nation, where most forests are reachable from seacoasts. The country's share of New Guinea is its last big concentration of primary forest in swaths large enough to qualify as intact forest landscapes. Most of Brazil's primary forest is over the 125,000-acre lower limit for intact forest landscapes thanks to the Amazon's immense, inaccessible geography.

Intact forest along the Sayan Mountains near Lake Baikal.

In 2020 another version of this quest for the great forests was published in *Science Advances.* Four dozen authors led by scientists from the Wildlife Conservation Society tagged the world's forests—pixel by pixel—high, medium, or low in terms of "healthiness." The rating of each pixel hinged on several factors: whether it contained farms, pavement, or some other substitute for natural habitat; the degree to which those disturbances were present on adjacent pixels; and how well connected each forest pixel was to other forest pixels. The study found that 40 percent of forests are in good health, and they are concentrated in five regions: the Russian boreal zone, the North American boreal, the Amazon, the Congo, and New Guinea.

We're not die-hard advocates of any of these systems. They all direct our attention to roughly the same places, showing us, according to their distinct methodologies, where the magic is still happening—where there

Rio Negro basin forests in Brazil, "primary" and intact. *(©Sebastião Salgado)*

are massive fully functional forest cores, which the planet needs to keep working.

Though we will focus on these five megaforests, there are other places we wish we could explore. Patagonia and Borneo would top a list of not-quite-so-mega, but still astonishing and important, forests. Patagonia's intact temperate forests are bounded by the Pacific Ocean in the west and the Andes mountain range in the east. The wet narrow strip in between, mostly in Chilean territory, grows splendid, healthy forests partially protected by a string of spectacular parks. Across the Pacific, the island of Borneo's forested center is still intact and fantastically diverse, including 90 percent of the planet's remaining orangutans. Malaysian and Indonesian settlements, oil palm plantations, and logging companies encircle the forest and have nowhere to go but inward. The island is on the brink of irreversible fragmentation—and is a cautionary tale for still-intact New Guinea.

Now let's leave the satellites to their orbits, descend through the clouds and canopy, and have a look around.

3

The North Woods

The Taiga

A journey into a megaforest often begins in a city such as Slyudyanka, which sits right at the pointy southwest end of Lake Baikal in south-central Siberia. Intact boreal forest rings the lake, and Slyudyanka, a stop on the Trans-Siberian Railway, is our gateway to millions of acres protected in the Tunkinsky National Park.

From the train station we strolled one late spring morning in 2019 through the town square, past a Soviet-era monument to the great railroad and found the stop where we could catch a bus into the park. The minibus pulled up right on time at 12:30. We climbed aboard and wedged into the back row of seats. The route wound through dense white birch forests over a low pass and into the Tunka Valley. Baikal, the world's deepest and largest freshwater lake by volume, fills a mile-deep (1.6-kilometer), bean-shaped rift that continues aboveground to the west, nudging the sawtooth Sayan Mountains north and the rounded Khamar-Daban range south into Mongolia. Lush forests blanket the valley slopes, eventually giving way to tree line and then snowy summits.

In front of us a blond giant scrolled through photos on his phone. John's 21-year-old daughter, Jessica, peered around the young man's branch-like arm intently enough that he noticed and struck up a conversation in Russian. He showed her some pictures that made her eyes go wide. There was one of a dead bear so big it looked fake, paws like baseball mitts. The young man, who identified himself as an elite paratrooper of some sort, said the bear had ripped six gold miners limb from limb in the woods north of Lake Baikal. Without being asked, he showed pictures of the evidence. The carnage was appalling and theatrical. Twenty minutes later we alit from the bus to explore the Siberian woods.

Sergei Natvevich, the Tunkinsky Region's top tourism official, met us. He quickly declared the bear story a fabrication. In his late 50s, the lifelong horseman is bowlegged and fit. He regularly participates in a competitive trail run up one of the lower Sayan peaks and fares well against much younger competition. He has a full head of silver hair and bronze skin over high cheekbones—movie-star good looks. He's an ethnic Buryat, a people whose ancestral territory surrounds Baikal. Today, the Buryats live mainly in the Russian administrative region of Buryatia, south and east of the lake, and in neighboring Mongolia.

"Have you ever been lost in the woods?" John asked Natvevich.

Natvevich, puzzled, looked in the rearview mirror at Jessica, who was translating from the back seat. He turned to look at the questioner in the passenger seat and then back at the translator. Natvevich turned down the Mongolian ballad playing on the radio of his Suzuki four-wheel drive. "What is the question?"

"Have you ever been lost in the woods?"

Natvevich burst out laughing. "I have never been lost in the woods, of course not! We grew up in the woods. It's our home." One hand on the wheel, he used the other to make a sweeping gesture at the larch, birch, Siberian pine, poplar, and spruce rushing by. Log houses with intricate scrollwork shutters dotted the valley. Natvevich turned up the radio again and sang along gorgeously. From the soaring emotion one would guess the song was about one of Chinggis* Khan's epic battles. In fact, it was about

* This is the updated spelling Mongolians and scholars use in reference to Genghis Khan.

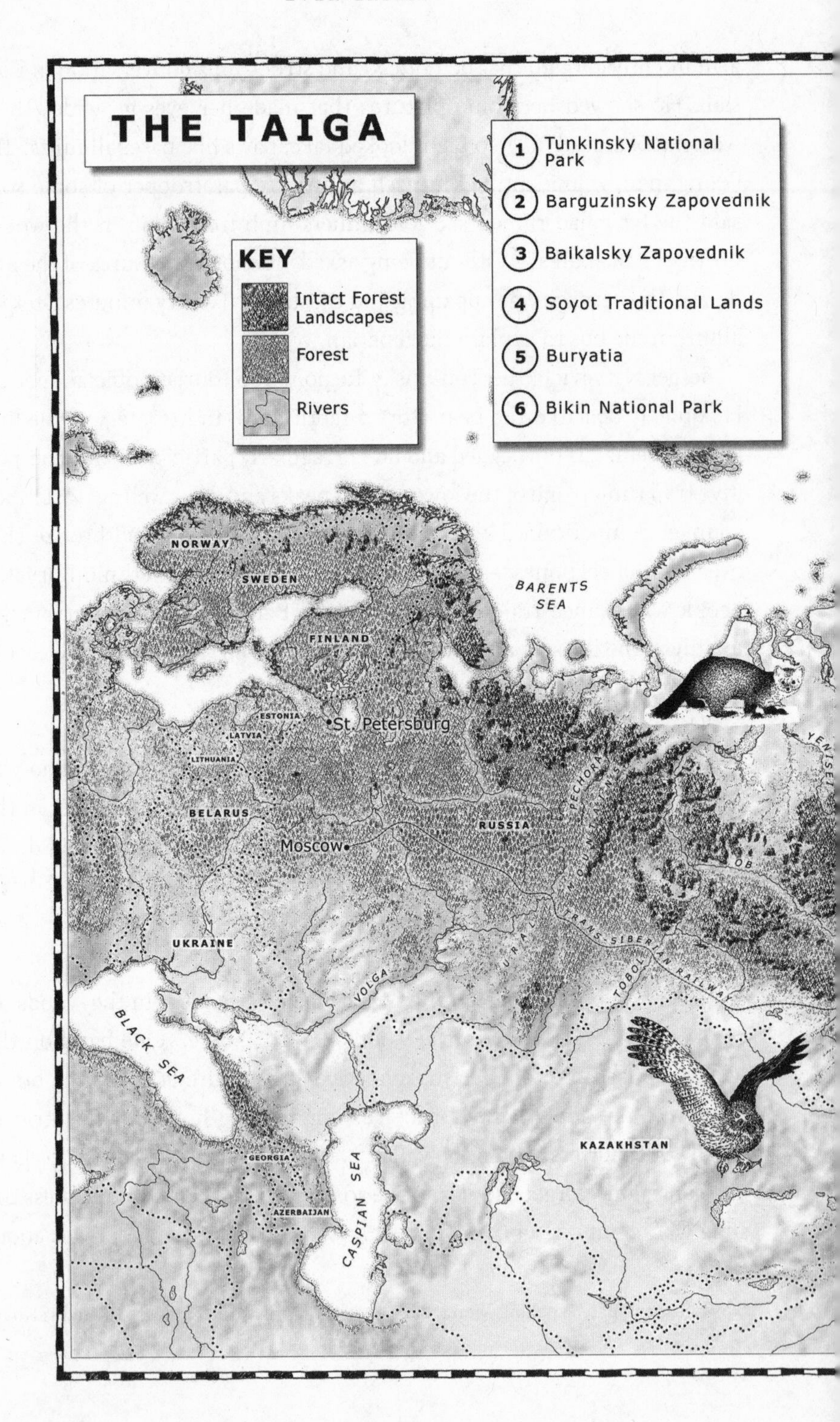
THE TAIGA
KEY
Intact Forest Landscapes
Forest
Rivers
1 Tunkinsky National Park
2 Barguzinsky Zapovednik
3 Baikalsky Zapovednik
4 Soyot Traditional Lands
5 Buryatia
6 Bikin National Park
NORWAY
SWEDEN
FINLAND
BARENTS SEA
ESTONIA
St. Petersburg
LATVIA
LITHUANIA
BELARUS
RUSSIA
Moscow
PECHORA
URAL MOUNTAINS
OB
YENISEI
TRANS-SIBERIAN RAILWAY
TOBOL
UKRAINE
VOLGA
BLACK SEA
KAZAKHSTAN
GEORGIA
CASPIAN SEA
AZERBAIJAN

ANGARA
BAIKAL
LAKE
SAYAN MOUNTAINS
Orlik
Irkutsk
Ulan-Ude
IRKUT
Slyudyanka
MONGOLIA
1
2
3
4
BERING SEA
KAMCHATKA PENINSULA
KOLYMA
LENA
Yakutsk
SIBERIA
AMUR
JAPAN
6
ANGARA
Irkutsk
5
CHINA
Vladivostok
MONGOLIA
N
S
E
W
750 MILES
750 KILOMETERS

Sergey Natvevich, a regional tourism official, at home in the woods of the Tunkinsky National Park.

how the singer's mom makes amazing tea. Natvevich said most Mongolian songs are about tea, parents, and horses.

Natvevich took us to the north side of the valley, where we walked a smooth trail through stands of powdery-soft white birch, whose coin-sized lime-green leaves winked in the sun. Siberian pines, called cedar by locals, threw sprays of long, dark needles among the birch boughs. In the fall, the conifers rain succulent pine nuts onto the forest floor. Dahurian rhododendrons speckled the understory with little magenta blossoms. Violets, columbine, Siberian wild iris, and anemone, in tiny whites and big fleshy yellows, dotted the ground. In mid-May the woods were coiled, ready to erupt with even more flowers, choke cherries, black and red currants, bilberries, blueberries, and ferns in the short, frenzied summer.

Snow had fallen briefly earlier in the morning, but now the woods were awash in sunlight and the temperatures climbed into T-shirt range. We nibbled wild onions, which, Natvevich pointed out, would rule out any romantic possibilities for the day. Then we sprawled on an inviting forest floor, springy with papery, snow-pressed leaves and moss, and gazed up at the pines. A cuckoo sounded up the slope in still-bare larches.

This most mega of megaforests is sometimes called the Eurasian boreal. The name is accurate, in that the forest spans those two landmasses, but obscures the fact that these woods are almost entirely in Russian territory. It takes two official continents to accommodate Russia and its forest. The ecosystem reaches into Scandinavia, but it is mostly fragmented there by roads and intensive timber production. We will generally call this forest the Russian boreal or just the Taiga. For clarity, we use "Taiga" with a big "T" to denote the name Russians give their entire boreal forest. We'll employ little "t" "taiga" as the scientific term that refers specifically to the most northerly or high-altitude fringe of all boreal forests, where they transition to tundra. The word originated in Mongolia, which sits at the southern boundary of the Taiga and in the Siberian Yakut language is synonymous with "untraversable forest."

Board a flight in Moscow bound for the Taiga's eastern extreme on the Kamchatka Peninsula and, weather permitting, you can spend the next 9 hours looking out the window at trees. Within Russia, the world's biggest forest blankets 2 billion acres, which is the size of the US lower forty-eight states. There are an additional 125 million acres in Scandinavia. The Taiga spans ten time zones.

The forest is mainly a mix of Scotch pine, Siberian pine, birch, fir, spruce, alder, poplar, aspen, and several species of larch. The proportions vary depending on latitude, elevation, soils, precipitation, aspect (which compass direction a hill faces), and the history of interaction with humans and fire. The far west, where Russia meets Scandinavia, is the only Taiga region that was fully glaciated in the Pleistocene. It is a birch-pine blend.

Snow-bent "drunken" birches in Baikalsky Nature Reserve.

Fast-growing white birches grow back first where forests have been logged or burned, which most have been in European Russia and western Siberia. Pine prevails along the Finland-Russia border, with spruce-fir evergreen forests over much of the remaining northwest and the Ural Mountains. Eastern Siberia has a swath of Scotch pine northwest of Baikal. Otherwise, larch reigns in the east.

The larch makes up over a third of the Taiga, or 655 million acres. Larches form the earth's biggest monodominant forests, areas in which the vast majority of trees are of a single tree species. The prolific larch (also known as the tamarack in North America) belongs to a tiny category:

deciduous conifers. Most of the year, larches show their crusty bark. Some angle their branches awkwardly as if to see something on the opposite side of the tree. Then June comes, and they sprout tiny bouquets of lime-green needles and live up to the tree's lovely Russian name, *listvenitsia*. The lacy emerald foliage lasts three months, then turns yellow and drops. Only twenty of Earth's 60,000 tree species grow needles and jettison them in the winter. One is the dawn redwood, thought to be extinct until a relict population was found in China in 1944; three are cypresses that grow in swamps. The other sixteen are larches.

Botanist Alexey Yaroshenko, now the director of Greenpeace-Russia, explains that needle dropping adapts the trees to life in the middle of the continent, where, far from the moderating influence of oceans, they are caught between permafrost and hot sun. Frozen ground denies the tree water. Needles, if the larch possessed them, would be crisped in winter by the intense sunshine. Larches also shrug off cold like no other tree on the planet, growing at up to 72 degrees latitude, farther north than the northernmost seacoast of Alaska.

The Siberian larch is rich in taxifolin, a member of the color- and aroma-giving flavonoid group that happens to be toxic to wood-eating fungi. That gives it exceptional rot resistance and explains why you can see larch houses and churches that have weathered centuries of Siberian winters. It's unruly to work with, however. Unless carefully cured, it splits and twists. Sergei Natvevich explained that the log house he was helping his son build would have larch for its lowest three or four rows, where ground moisture can rot the walls. The rest would be made of the more easygoing pine.

The Pacific fringe of Russia's great forest is its biodiversity hot spot. Warmer conditions, particularly in the far southeast, produce much more varied trees, including oak, maple, yew, linden, ash, black birch, and poplar. Amur tigers and leopards roam forests that have fall foliage like New England's and salmon streams like those in the Pacific Northwest. Among the spawners are all six species of Pacific salmon, some of which feed Blakiston's fish owl, the world's largest. Unlike most owls, which have silencing adaptations on their flight feathers and acute hearing for catching terrestrial critters, fish owls are noisy fliers and don't hear particularly well. That's okay, because the fish and owls can't hear each other through the water. The birds can survive blisteringly cold winters and judge the position of their prey through the distorting lens of water. Only a mature forest can provide the 2-foot-tall raptors with trees big enough for their nest holes.

The species-rich corner of the far southeastern Taiga was made famous among Russians by the early twentieth-century army captain Vladimir Arsenyev. He turned the notes from his survey missions into popular accounts, taking some liberties with story line but reporting in precise detail the flora, fauna, geography, geology, and weather, as well as his interactions with the Chinese, Korean, Russian, and Indigenous peoples who hunted, trapped, and collected ginseng there. A soldier-naturalist in the mold of Meriwether Lewis, Arsenyev lionized his Indigenous companion, Dersu Uzala, the title character of a dreamy 1975 Akira Kurosawa film. Like Lewis's guide Sacagawea, Arsenyev's Native companion bailed the explorer out of more than a few tight spots.

Twenty-five percent of Russia's forest zone is made up of wooded ecosystems that meet the IFL minimum of 125,000 acres. That's down 10 percent from the year 2000. In the northern Taiga, the loss was attributed 56 percent to fire and 41 percent to oil, gas, and mining, with the tiny remainder due to logging. In the south, with larger, more accessible trees closer to the colossal Chinese market, logging accounted for 54 percent of losses; 23 percent was due to energy and mining and 21 percent to fire. Some fires are caused by lightning, but the Russian government and Greenpeace concur that the great majority—90 percent, according to the NGO's mapping team in Moscow—is started by people. The map analysts arrived at that conclusion by combining NASA's spatial data on new fires with high-resolution maps of roads and other markers of human presence.

In mid-August 2019, the Russian Federal Forestry Agency reported 5.9 million acres of active fire in its eastern forests. That's around the average for the whole Siberian fire season and a million acres more than burned in California's record-shattering 2020. All burning at once. During our 2019 trip to the region, we drove through unprotected forests at the beginning of that Siberian summer of infernos, coming across an area that went up in a 2016 conflagration. Three years later it was still ghastly. The remaining stalks of pine and birch stood defoliated on exposed hillsides, like something we shouldn't have been seeing. A month before our drive, a forest fire had vaporized a village several hours to the east. Forestry officials then put big swaths of Siberia's woodlands off-limits. We passed hundreds of candy-striped barriers used to implement this policy. The flimsy thigh-high poles blocked entry to every dirt side road. Signs announced a 4,000-ruble fine for simply entering the woods, independent of whether one just walked around, lit a campfire, unleashed a dog, fired up an ATV, harassed the wildlife, or poached a tree. The $62 fine was roughly 20 percent of monthly per capita income in Buryatia. When we asked a local Buryat friend whether the barriers work, he laughed: "The only people this keeps out are the kind of people who obey the law."

Oil and gas exploration are persistent and multifaceted forest threats in the Taiga. "Especially in Siberia, and especially in the area of the pipeline that now goes to China," said Peter Potapov in a video interview, referring to a link inaugurated in 2011. "I think that is the driving force. We see the expansion of this network of the seismic lines, then they install the pumps and they put in the pipelines and roads. As they go south, closer to the road infrastructure and closer to the consumers, you can see logging goes side by side with this expansion. So, logging will start from these roads and go inside the forest. But the main effect, of course, is fire. A lot of fire starts from the infrastructure. It doesn't mean that the people who do the exploration specifically will burn forest. It can be hunters who use the roads later who will set the fires accidentally." Potapov added that mining, particularly for gold, is locally devastating to forests, waterways, and wildlife.

The Russian Taiga is home to outlandish fauna, including the world's biggest tigers and the tiny, fanged Siberian musk deer on which the cats are known to feed. It has the world's largest salmonid, the taimen, which can grow to 6 feet in length, weigh over 100 pounds, and live to the age of 55. Alongside these unique creatures are forest animals familiar to North Americans. Kamchatka has one of the world's largest concentrations of brown (grizzly) bears. Russia's woods have elk, lynx, moose, wolverine, bison, and caribou, which are called reindeer in Eurasia. These shared species wander among trees also familiar in North America—pine, fir, spruce, birch, aspen, poplar, and larch.

The Taiga and North American boreal forest have similar fauna and trees because they were recently one circumpolar super-megaforest interrupted only by the North Atlantic. Before melting ice filled the Bering Sea, about 11,000 years ago, you could walk from Norway to Newfoundland. Moose and people walked to the Americas, while horses and bison went from Alaska to Russia. Mammoths are buried in permafrost on both sides of the strait. During the last period of intense glaciation, an ice-free zone known as Beringia connected present-day Russia, Alaska, and the Yukon in a huge grassy refuge roamed by the connected continents' herbivores for millennia.

The North American Megaforest

The North American megaforest spans 1.5 billion acres at the top of the continent. Not the very top; that's tundra, a frosty heath and grassland that picks up where the climate is too cold and dry for tall plants to develop. The big forest of the US and Canadian north covers a region nearly nine times the size of Texas, marching through central Alaska and across Canada, cradling Hudson Bay and reaching to the shores of the Atlantic. The North American megaforest also has a temperate edge that lines the Pacific Ocean along the coast of Alaska and British Columbia, a conifer-heavy counterpart to the diverse salmon forests of the Russian Far East. The ocean buffers this part of the forest from extreme temperatures and sends brimming rain clouds ashore. That creates the planet's largest temperate rainforest, 50 million acres of the United States and Canada.

The northern reaches of the North American forest are the most intact woods in the world. Nearly 80 percent qualify as intact forest landscapes, of which just 3.3 percent were fragmented or lost from 2000 to 2013, a third of the global loss rate during the same period. If we consider Canada as a whole, including its southern cities and the country's commercial timberland, we find a forest that was still 51 percent IFL in 2016, with a 5.8 percent loss since 2000.

There's no single word for "forest" in the language of the Kaska Dena, whose traditional territory spans some 60 million acres of largely intact woods in the Yukon, British Columbia, and Northwest Territories. One expression is *ts'udón*, which translates roughly as "densely spruced." This makes sense when one visits the Yukon; the world seems to be made of white spruce. These trees coat the lower slopes of snow-topped mountains and carpet valley floors. Unlike the loose-limbed swaying lodgepole pines one sees farther south, spruce twitch in the wind. Some are so narrow and nubby they resemble pipe cleaners. Others are elegant cones like Gothic cathedral spires. In late October of 2019 in the southern Yukon, the spruce tops were brown. They appeared to be perishing from a blight but on closer inspection were revealed to be lavishly hung with soft cones. It was a mast year, meaning that the trees were opting, as a group, to cash in huge stores of carbohydrates and make a crazy number of seeds, so many that the seed-eating squirrels and voles would be unable to keep up. That leaves seeds in their trillions to sprout new spruces. Historically this happens around every ten to twelve years and is not a tree-by-tree decision. To overwhelm seed predators, trees must mast at the same time. Environmental conditions alone are an incomplete explanation for the remarkable synchronization, as stands up to 1,500 miles apart will mast in the same year. In forests farther south in British Columbia, trees have been shown to communicate through fungus enmeshed in roots, which plug into what's called a mycelial network, referring to threadlike fungal connectors.

At around 61 degrees north latitude, 5 degrees shy of the Arctic Circle, you can find the northernmost office of the Wildlife Conservation Society (WCS). The building is an unassuming one-story affair shared with Ducks Unlimited Canada in a light-industry district of Whitehorse, the capital of Canada's Yukon Territory, which neighbors Alaska. A few yards away, the Yukon River steamed in the sunshine of the chilly October morning when we arrived, doffed hats and jackets, and accepted some coffee. Despite the billion and a half acres of surrounding woods, if you walk into

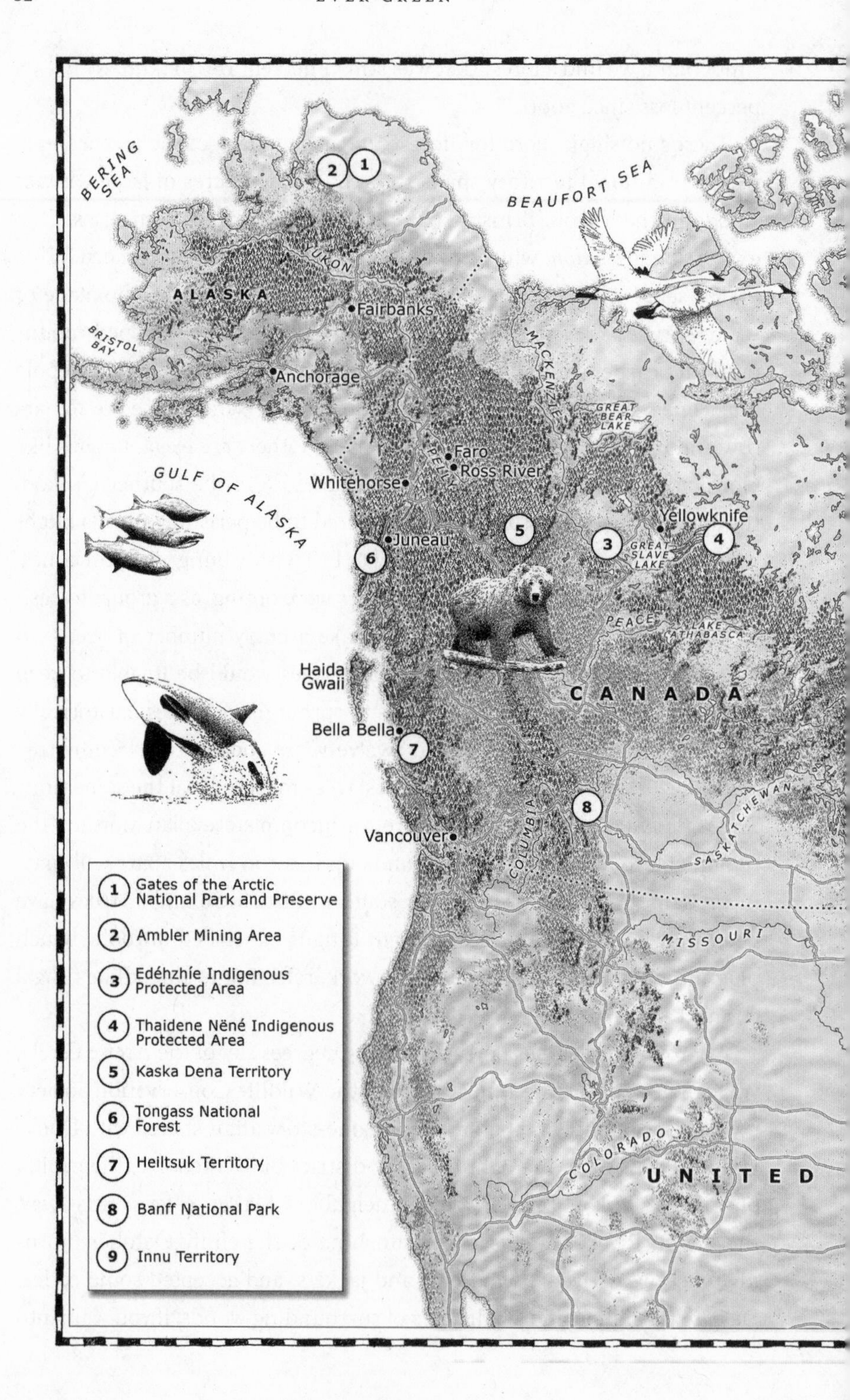
BERING SEA
BEAUFORT SEA
YUKON
ALASKA
Fairbanks
BRISTOL BAY
Anchorage
MACKENZIE
GREAT BEAR LAKE
Faro
PELLY
Ross River
GULF OF ALASKA
Whitehorse
Yellowknife
Juneau
GREAT SLAVE LAKE
PEACE
LAKE ATHABASCA
Haida Gwaii
CANADA
Bella Bella
SASKATCHEWAN
Vancouver
COLUMBIA
SASKATCHEWAN
MISSOURI
COLORADO
UNITED
1 Gates of the Arctic National Park and Preserve
2 Ambler Mining Area
3 Edéhzhíe Indigenous Protected Area
4 Thaidene Nëné Indigenous Protected Area
5 Kaska Dena Territory
6 Tongass National Forest
7 Heiltsuk Territory
8 Banff National Park
9 Innu Territory

THE NORTH AMERICAN MEGAFOREST
LABRADOR SEA
HUDSON BAY
KOKSOAK
9
Quebec City
Montreal
Ottawa
ST. LAWRENCE
LAKE SUPERIOR
LAKE HURON
LAKE ONTARIO
LAKE MICHIGAN
LAKE ERIE
ATLANTIC OCEAN
TATES
MISSISSIPPI
KEY
Intact Forest Landscapes
Forest
Roads
Rivers
500 MILES
800 KILOMETERS

the WCS-DUC office hoping to discuss trees, you'll need to be patient. First the scientists there need to talk about water, fire, and ice.

Because this megaforest's story is really two stories. The first is about glaciers. In recent geologic time, the Laurentide and Cordilleran ice sheets crept back and forth across a large part of North America. Glaciers reached their southern limits at Indiana between 25,000 and 26,000 years ago and began to retreat 10,000 years later. Coming and going, they raked a labyrinth of lakes, ponds, fens, swamps, marshes, and rivers. That's why Ducks Unlimited is here; the venerable group of hunter-environmentalists protects the wet habitats of ducks and geese.

DUC's soft-spoken Jamie Kenyon explained that we were in a very wet forest. Around James Bay and Hudson Bay, 75 to 100 percent of the "land" is, in fact, aquatic. One expects to find wetlands mainly in flat places because gravity drains water from inclines. But here even the hills can be spongy. One day, as Kenyon and a colleague were laboring hand over hand up a precipitous slope, they happened on a tiny spruce and orange sphagnum moss, both organisms they would expect to find in wetlands. They dug into the hill and found a thick stratum of carbon-rich peat—the two were scaling a rare "slope bog."

All the still water is great if you're a larval insect. When it thaws in spring, a gargantuan cloud of six-legged biomass launches. Birds travel from the United States' lower forty-eight, the Caribbean, Central America, and South America to gorge and breed. It's worth the trip. The 1 to 3 billion migrants increase their numbers to 3 to 5 billion over the summer with the addition of brand-new, well-fed warblers, vireos, finches, tanagers, thrushes, flycatchers, ducks, geese, and swans, among other species, that disperse to the wilds—and backyards—of the warmer parts of the hemisphere each fall. A billion boreal birds winter in the continental United States alone.

Fire is the region's other story, taken up by Don Reid and Hilary Cooke of the WCS. Tall with long hands and wire-rimmed glasses, Reid is the office elder, kind and calm. It was relaxing just to be around him. Dark-haired and freckled, Cooke is Reid's energetic opposite. "I'm passionate about woodpeckers," she enthused about the birds on which she wrote her

PhD. But Cooke seemed passionate about every aspect of the megaforest outside the building. Including fire.

They explained that if you could put the boreal forest on fast rewind, you would see fire hopscotching around the landscape, with any given region burning down every 100 to 200 years, depending on the region. This forest has been forged by blazes. Reid and Cooke have a bone to pick, very politely, with the global IFL maps, on which fires can be readily confused with clear-cuts. That knocks large areas out of intact forest status and creates the dispiriting impression that the boreal is in worse shape than it really is. If one deducts fire from IFL losses (which Potapov and colleagues now helpfully do), the losses drop by over 70 percent. Lightning sets most of the fires, which, over many decades, create a patchwork of forest at various stages of regrowth.

Right after a fire, forests are alive with exquisitely evolved pyrophilous (fire-loving) insects. The black fire beetle, for example, uses infrared detectors in its thorax to find burned trees. The white-spotted sawyer can smell volatile chemicals called monoterpenes, which are produced in large quantities by dying trees and transported far and wide by smoke. These insects respond to fire immediately and can arrive and start dining on scorched wood even before the flames are extinguished. In turn, the beetles feed black-backed, three-toed, and hairy woodpeckers, which nest in spruce trees killed by fire. The blackpoll warbler preys on the fire bugs before migrating from one megaforest to another—the Amazon—where it winters along streams among parrots and toucans. At the other end of the burn spectrum are the most mature woods, thick with spruce and pine and carpeted in lichens that nourish caribou. Every stage in between sustains its own collection of plants and animals.

The snowshoe hare, for example. That's Don Reid's specialty. The animals have wide hind paws that spread their weight on the snow and a pelt that toggles between winter white and summer gray-brown. To study them, Reid collects droppings, the ebb and flow of which track rabbit population cycles. Hare numbers explode and plummet every eight to ten years. First the animals breed like rabbits and reach a peak population, which subsequently collapses to as little as 1 percent of its maximum.

Wildlife Conservation Society biologist Don Reid in the Fox Lake burn.

Droppings also tell Reid where the hares thrive. When the scientist offered to take us to see their forest habitat, what he had in mind were burns.

We drove an hour north of Whitehorse through dense spruce forest until, quite suddenly, the woods lost their color. Reid pulled off onto a frozen dirt track in a spectral, dreamy landscape of young trees rimed with ice so that they seemed to produce their own light. Bare aspens predominated, supervising a few spruces the size of tabletop Christmas trees. Reid parked and led the way up a hill covered in just enough snow for tracking. We were in the Fox Lake burn.

It took us a while to find a set of tracks. The four pawprints were bunched together as if their owner had been pogo-sticking with its limbs bound together. The rear paws were slightly in front of the front ones. You have to imagine, Reid explained, the hare bounding away from a lynx. The brawny back legs launch the bunny off of its snowshoes, which land ahead of the dainty forepaws just as these lift off for the next bounce. So the four impressions end up clustered together with the back in front. Reid shook his head at the scarcity of tracks. This forest wasn't ready for rabbits yet.

Animals seek habitats with two qualities: a lot of food and a low probability of becoming food. The snowshoe hare thrives in places with tender leaves and shoots growing no more than 3 or 4 feet off the ground; it can balance on hind paws, and snow provides a boost in winter. But hares also need tree cover, ideally conifers, to obscure the view of predators, including great horned owls, lynxes, martins, northern goshawks, and coyotes. Places burned by big fires are perfect fifteen to fifty years after the blaze. At that stage, the bigger spruces and aspens block raptors' lines of sight. Shrubs obstruct the cat-, dog-, and weasel-family members. And yet, there is still plenty of tender knee-high aspen, willow, pine, and spruce to eat. By contrast, in smaller clearings created by ubiquitous logging farther south, owls and hawks perch on tall trees at the patches' edges and wait for passing rabbits. The Fox Lake burn took out 111,459 forest acres in 1998. Twenty-one years later, it offered the bunnies copious food. But the probability of becoming food was still too high, because deciduous aspens don't give the hares enough cover, and the conifers are returning at a slower-than-normal pace. Hotter, drier summers could be a factor, Reid said.

For purposes of comparison, we retraced our route a few miles to the south and entered mature woods that had escaped the fire. There was no straight path through the spruces. Weaving was easy enough, with a certain amount of pushing lower branches out of the way. Afternoon light slanted through branches over a slowly freezing creek. It was quiet except for scolding red squirrels. Roses with scarlet hips decorated the understory, where we also found cone caches organized by the squirrels. There is evidence of squirrel foresight in the location of the piled cones: they choose branches that will make telltale snow fins over the food in the middle of winter.

In spots, our feet sank 6 inches through a crust of moss. At first it would resist ever so slightly and then give way. Reid reached down into his own bootprint, his arm disappearing halfway to the elbow, and pulled out a sprig of moss. Given how delicately the moss crunched and the brevity of the growing season, we guessed that it takes a long time to heal the print. Reid confirmed this but did not seem overly concerned about the damage our brief meander was doing to the 1.5-billion-acre forest. The old stand of spruce was lovely for humans to walk in and apparently a bonanza for squirrels, but it was too old for hares. New tree growth was out of rabbit reach, and the canopy shaded out other foods, such as willows. There was plenty of cover from the owls, but the rabbits couldn't get a meal. Without fire, bunny numbers would be drastically reduced across the boreal landscape.

As climate change accelerates, however, the story of burns is becoming more complicated. Since the 1960s or 1970s, fires in Canada have been increasing. One estimate is that from 1959 through 2014, each year fires burned 82,000 more acres than the year before. Fire season now starts a week earlier and ends a week later. The trends vary across the country, with relatively little change in the moister east and dramatic fire expansion in the inland west. Where fire comes more frequently, caribou go hungry because the lichens they favor grow in old forests. Fire revisits some forests before trees can attain sufficient girth to accommodate woodpecker nests. Younger trees have fewer cones, so there's less food for seed-eating squirrels and voles, which means fewer of these furry creatures available for predators, smaller deposits made to the "seed bank" held in the soil, and, consequently, scarcer spruce sprouts. One black spruce wood in the Yukon that has historically burned every 80 to 150 years went up in 1991 and again 14 years later. Spruce is adapted to regrow in its own ashes, but the trees regrowing after the first blaze were too young to provide seeds following the second one.

Too-hot fires can burrow into peat and stay underground in a sort of pyro-hibernation before resurfacing in spring along with the bears. Or they can incinerate seeds in the soil and move on to scorch the formidable aspen roots that otherwise would produce new trees after a "normal"

Spruce cones cached by red squirrels in a mature Yukon forest.

fire. Today's fires can thus scald forest into a shrubbery of willows, whose small seeds, equipped with downy sails, can disperse on the wind without the help of animals.

Back at the office, Hilary Cooke pulled out a map showing what kind of nature will prevail in the Yukon over the next century. Until recently, vegetation types have been called biomes, a contraction of the phrase "biotic communities of Möbius," homage to a nineteenth-century German zoologist. Biome classification is based on vegetation. These classifications give the impression that the plants found in a place are constant, like geology. In the boreal zone, where climate change is galloping, biome labels can no longer be written in permanent ink. The map Cooke displayed uses a new word, "cliomes," coined during a joint climate planning effort between Alaskan and Canadian researchers. Because the climate changes, cliomes move. By the 2090s, trees will infiltrate tundra in the north. Grasslands may supplant forest in the south. The Yukon could lose seven of eighteen cliomes and gain one it didn't previously have. Some places that have been stable for centuries are predicted to experience multiple cliome flips over the course of this one.

In this new reality, conservationists can't hope to preserve places in amber. "What we're protecting is the stage, not the actors," said Cooke. "Protecting the stage" comes from the title of a 2010 *PLOS ONE* paper by The Nature Conservancy's Mark Anderson and Charles Feree, who say societies should let nature respond to climate change by keeping diverse geologic and physical spaces undeveloped rather than trying to secure every species in its current habitat. Some will perish, some adapt and stay, and yet others may move—perhaps even with a human assist—and thrive on new stages. In a larger sense, the whole megaforest, wild and changing, is a stage nature needs to adapt to the volatile climate.

After climate change, mining is arguably the most widespread threat to the North American boreal. From above, the Faro Mine looks like someone dropped two bombs on the forest, 10 miles apart, and then built a road between the craters, adding subdivisions and a golf course nearby. Faro is 4 hours north of Whitehorse on the Pelly River in the traditional territory of the Kaska Dena people, who invited us to learn about their plan to create a very large protected area. A big part of their plan involves zoning their ancestral lands to confine mining to areas where it poses less risk to the ecosystems they depend on.

Faro was once the largest open-pit lead-zinc mine in the world. It operated from 1969 to 1997 and is now Canada's number two toxic cleanup site by cost, thanks to 70 million metric tons of tailings, the material that's had valuable minerals extracted from it. There's another 320 million tons of waste rock that was excavated to get at the ore. Dealing with those materials is set to begin in 2022 and require $450–$500 million (Canadian), on top of $500 million already spent to avert environmental

White spruce rings Swim Lake in Kaska Dena territory. The lake lies atop an unmined deposit of zinc and lead.

disaster in the meantime. The mine's operator, Anvil Range Mining Corporation, went bankrupt upon closure and stuck the Canadian taxpayer with the tab.

It might have been worse. Kaska elders John Acklack and Clifford McLeod guided us into the bush on four wheelers along old mining exploration trails to see a spot the miners originally considered digging. We maneuvered along snowy paths for a couple of hours and pulled up at Swim Lake, where wavelets rattled ice on shores ringed by white spruce. This is a spot where their families traditionally camped for a couple of weeks each year to catch spawning fish called suckers. In the 1960s, the two worked here at a camp for a prospecting crew. Ore was identified under the lake, which was spared when material was found in a drier spot closer to the main road.

Firms that dig up copper, gold, zinc, lead, molybdenum, and other metals in the north argue that mines have small footprints. The footprint metaphor suggests something small and harmless. To understand the possible impact of mines, one needs to imagine footprints traipsing in all directions—especially downstream—from the place where the digging happens. Remember that the boreal forest is a forest of water, shot through with creeks, rivers, lakes, fens, swamps, bogs, marshes, and ponds. That is why Faro is a disaster. Water passing through mine tailings and waste rock becomes highly acidic and can leach toxic metals into rivers, lakes, and groundwater. Faro's tailings are tenuously held in a 5-mile dammed section of Rose Creek at a cost of around $40 million per year. Even while this dam is intact, the creeks most directly exposed to the mine have in recent years registered manganese levels up to twenty times government standards, iron seventy-five times too high, and zinc as much as fifteen times the level considered safe. If the containment were to fail—as two mining dams did quite spectacularly in Brazil in 2015 and 2019—the nearby Pelly River, which flows into the Yukon River, would be in peril, putting salmon, ducks, bears, and people at risk.

The proposed Pebble Mine in Alaska, fortunately shelved for now, would pose a similar risk to a glaciated landscape of creeks, ponds, wetlands, and forests. The water flows into Bristol Bay, home of the largest wild salmon fishery in the world. Fishermen and Native leaders protested this gold-copper dig on state land, saying it would menace rivers and a bay frothy with sockeye salmon; 58 million fish returned in 2020. The mine would also require roads, which beget more prospecting and greater peril to the fish. In September 2020, undercover environmentalists recorded Pebble executives saying that the mine footprint for which they were seeking approval would be just the beginning of a complex up to nine times bigger. The project was rejected by the Army Corps of Engineers in November of 2020.

Roads to mine sites are called "resource roads" in this part of the world. The term implies the narrow purpose of getting a resource out. So limited is the public road network, however, that driveways of 60 miles or more are needed. In practice, these single-purpose spurs serve multiple users, branching and radiating into a web of smaller dirt roads and then tracks passable by snowmobiles and four wheelers. Hunters can get within an easy tromp of every caribou and moose in the forest. An area with this kind of infrastructure holds little refuge for big game, which, the Kaska complain, are being overharvested by outsiders around Faro and in other parts of their territory similarly roaded.

Seismic lines are a related problem. They are cleared strips along which oil companies detonate rows of explosives. Echoing shock waves tell where there's oil and gas. Peter Potapov recalled his puzzlement when first analyzing satellite images of Canada. "You can see exploration that happened decades ago. It looks like a grid laid over the forest. When we first saw such images, we thought, 'something wrong has happened with the satellites!'" But the grids are real. There are over a million miles of seismic lines in Alberta. In the neighboring Northwest Territories, "seismic lines are the single largest landscape disturbance caused by humans," according to the territorial government.

This disturbance is particularly bad for woodland caribou, which are the canaries in the coal mine of the boreal fragmentation. Caribou are Canadian icons, featured on the quarter coin, and central to traditions and identity for various First Nations of the north. In this forest of water, caribou have thrived in part due to a special adaptation, hollow fur follicles, that makes them float better. Caribou can swim across big lakes and rivers on their migrations and wade deep into bogs, fens, swamps, and other wetlands that sparkle across the boreal. The buoyant ungulates feed on the tender shoots of water plants.

The best-known and more readily seen "barren-ground" caribou travel in great herds across the Arctic tundra and may winter in boreal forest. Woodland caribou, on the other hand, are fully forest creatures and vulnerable to forest disturbance. They live entirely in the woods and adjacent mountains, moving frequently so as not to overgraze lichens and to stay ahead of wolves. Thirty-six of Canada's fifty-one boreal herds are in big trouble. Roads and seismic lines make both wolves and human hunters more lethal. They travel with greater ease in forests thus fragmented and penetrate caribou redoubts that were previously safe. Furthermore, bushes and grasses that sprout in linear clearings draw deer and moose, creating a critical mass of prey animals, an irresistible buffet that draws wolf attention to areas where woodland caribou would otherwise go unnoticed.

The caribou's early response to North American boreal forest fragmentation is a warning about carbon gases that, through the workings of climate change, can prove just as deadly as carbon monoxide in coal shafts. Globally, there are 190 years' worth of worldwide emissions tucked away under boreal forests. That buried carbon is like a slumbering beast whose sensitivity to the disturbances on the surface we don't fully understand. Creating forest edges with seismic grids, roads, power lines, mines, pipelines, and logging in a drying climate warms the forest floor, making fire likelier and soils quicker to release CO_2. The precise degree to which these changes will destabilize the carbon under the North American boreal is unclear. We haven't had time to find out. But maybe the decline of the caribou is telling us all we need to know about the tolerance of this biophysical system to disturbance. It's an experiment we only get to run once. How much do we want to poke the beast?

4

The Jungles

The Amazon

The Amazon's answer to the big, water-loving caribou is undoubtedly the tapir. The size of a Shetland pony, this herbivore is the biggest land animal in the Amazon and an ancient relative of the rhinoceros. It is equipped with a short mane and droopy prehensile trunk between its watery eyes. And, despite a barrel-like physique, the tapir steps mincingly. One early October evening by a river in a remote corner of the Amazon, one of these beasts walks out of the forest and onto the beach. Her hooves are mute in the sand, splayed dewclaws leaving birdlike tracks. She moves to a soundtrack of parrots' and macaws' abrasive squawks, trilling songbirds, cicadas droning, and turkey-like curassows, which emit a hollow thump. The primate section is a counterpoint of howler monkeys' stirring roar and orgasmic squeals produced by titi monkeys. Our tapir wades in and swims the channel silently.

This stretch of the Madidi River in Bolivia is a good place to see a tapir. It's as intact as tropical forest gets. Its native pigs troop about in the hundreds, and huge patterned catfish swim the green waters. Emergent trees, logged elsewhere, spread crowns like lookouts high above the main canopy. This spot is in the Amazon's exuberantly vital western reaches,

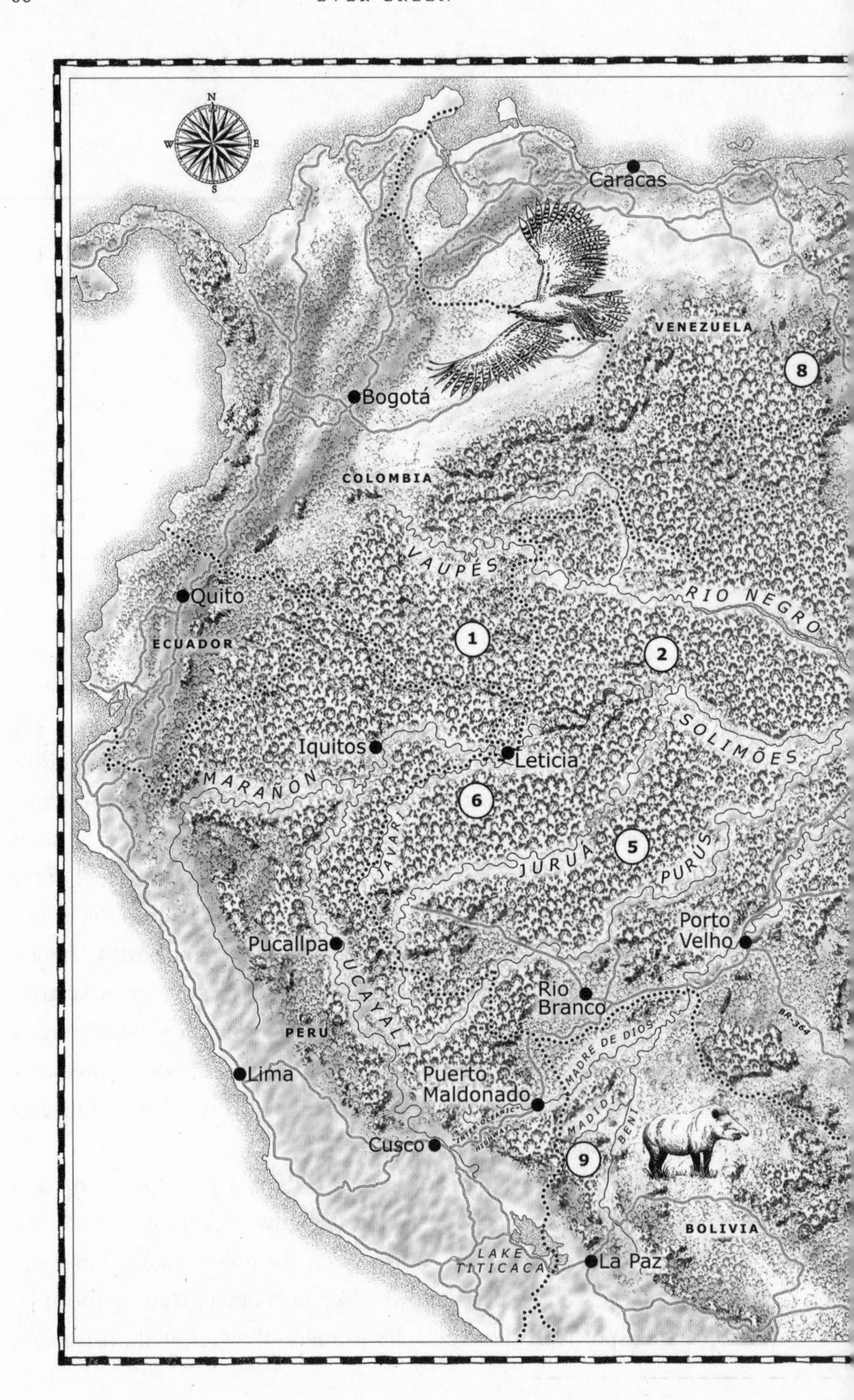
N
W
E
S
Caracas
VENEZUELA
8
Bogotá
COLOMBIA
VAUPÉS
RIO NEGRO
Quito
ECUADOR
1
2
SOLIMÕES
Iquitos
Leticia
MARAÑON
6
JAVARI
5
JURUÁ
PURUS
Porto
Velho
Pucallpa
UCAYALI
Rio
Branco
BR-364
MADRE DE DIOS
PERU
Lima
Puerto
Maldonado
MADIDI
BENI
Cusco
9
BOLIVIA
LAKE
TITICACA
La Paz

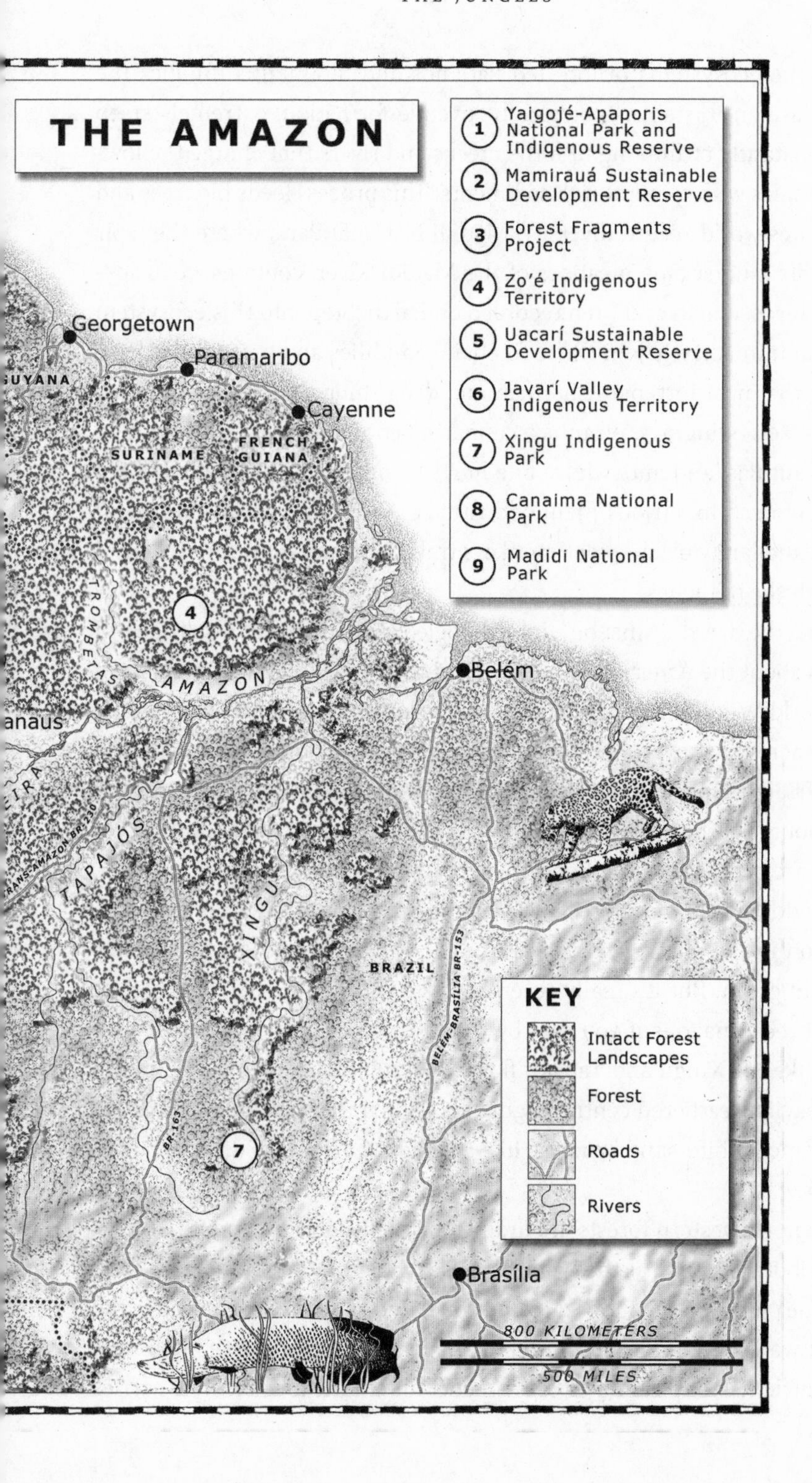
THE AMAZON
1 Yaigojé-Apaporis National Park and Indigenous Reserve
2 Mamirauá Sustainable Development Reserve
3 Forest Fragments Project
4 Zo'é Indigenous Territory
5 Uacarí Sustainable Development Reserve
6 Javarí Valley Indigenous Territory
7 Xingu Indigenous Park
8 Canaima National Park
9 Madidi National Park
Georgetown
Paramaribo
Cayenne
SURINAME
FRENCH GUIANA
TROMBETAS
AMAZON
Belém
TAPAJÓS
XINGU
BRAZIL
BELÉM-BRASÍLIA BR-153
BR-163
Brasília
KEY
Intact Forest Landscapes
Forest
Roads
Rivers
800 KILOMETERS
500 MILES

where the ecosystems of forested flatlands and mountains mingle. The Andes are an unstable, geologically juvenile formation, extremely steep and constantly crumbling into the creeks and rivers that charge the lowland jungles with mineral-rich sediments. This process feeds big trees and stimulates world-record diversity. Madidi National Park, where the tapir frequents a horseshoe meander of the Madidi River, contains 1,028 species of birds, one in every ten recorded on Earth. Step into this ecosystem and you immediately join the food chain. Sandflies and mosquitoes keep all but the most intrepid fishermen and drug smugglers away. There are electric eels, stingrays, piranhas, caimans, venomous snakes, wasps that kill tarantulas, and ants whose bite hurts so much they are called bullet ants and used in various manhood rituals. And there are jaguars, giant otters, and lanky black spider monkeys traveling hand over tail over hand through the branches.

Type the word "Amazon" into Google, click through five pages of results about the American company, and then you will find an item from the World Wildlife Fund about the world's largest rainforest. It covers 1.4 billion acres, just over 1 percent of the planet's land, and contains 10 percent of Earth's species. The central waterway of the Amazon basin is 4,100 miles long, nearly two Mississippis. It's fed by over 1,000 tributaries, 17 of which are over 1,000 miles long.

As with all the megaforests, it would be more accurate to refer to the Amazon in the plural: Amazons. From above, the forest is a deceptively uniform green. But it's really several forests within a forest. There are at least three Amazons if you sort by the color of rivers. Transparent blue rivers like the Xingu and Tapajós flow south to north, down from Brazil's old, heavily weathered central highlands. When the Tapajós's level drops in summer, white-sand beaches, like a saltless Caribbean, draw locals to the water.

The northwestern forests are drained by the Rio Negro and its Colombian tributaries, such as the Vaupés. These rivers run the color of black tea. Take the plunge in the Rio Negro and you'll find yourself luxuriating in the softest water (minerally speaking) of any major river on Earth, with relatively little risk of piranha or caiman encounter. The low-silt river carries

Tapir by the Madidi River.

organic acids leached from leaves on sandy forest floors. High acidity and low nutrients mean, among other things, fewer biting insects; we've slept in the Rio Negro jungle without the mosquito nets that are frequently required to avoid being carried off in the night by Amazon insect life. Low-bug forests mean leaner food chains and smaller fish in the marketplace. Forest soils are so thin in places that the tree roots form a springy mat over the rocks. The environmental upside of having almost no soil is that the Rio Negro region has attracted little interest from agribusiness and remains predominantly under the legal control of Indigenous societies, which have thrived by enriching small jungle plots into deep "black gold" soils through centuries of composting and careful cultivation.

The main stem of the Amazon and its western tributaries are the "white" rivers, in which, for the most part, you won't want to swim. From October through March, torrential rains pummel the Andean headwaters, turning the waterways into a rich soup of sediments. The "white" rivers of the Amazon come in a range of corals and browns, depending on the mineral composition of their silt. They flood annually, fertilizing a forest that is rich in everything: fruits, insects, bats, birds, monkeys, snakes, frogs, cats, dolphins, and fish. Fish leap to pick fruit straight from branches. Nearly blind pink dolphins have highly flexible spines for weaving among the submerged trees to hunt. In contrast to most of the world's dolphins, their neck vertebrae are unfused, like our own, an adaptation that enables head swiveling. And there are, as we saw in Madidi, things that may bite or sting a naive visitor taking a dip to cool off in the equatorial heat. Crocodilians, for example, such as the black and spectacled caimans.

Calling the 4,100-mile main stem Amazon "the Amazon" oversimplifies the question of tributaries. It also sidesteps a difference of opinion between Brazilians and Andeans about where the name should start being applied. The tributaries most distant from the Amazon's mouth are the snow-fed Peruvian Mantaro and Apurimac Rivers, which fuse into the Ene, which becomes the Tambo, before flowing into the Ucayali, which makes a 907-mile northbound run. Around 70 miles upstream of Iquitos, a city of some 500,000 people, this river joins another Peruvian monster

The "white" water Juruá River in Brazil. *(©Sebastião Salgado)*

current, the Marañón. That confluence, say Peruvians and the people of other Spanish-speaking countries in the vicinity, is the beginning of the Rio Amazonas. When it hits the Brazilian border 391 miles downstream, however, the receiving nation rebaptizes it the Solimões, the name under which it travels for the next 1,056 miles. At Manaus, the murky Solimões absorbs the Rio Negro and becomes the river Brazilians know as *O Rio Amazonas* for the rest of its journey to the Atlantic Ocean. The river was named after the Greek female warriors because Spanish explorer Francisco de Orellana described battling very tall, fierce women at a spot near the mouth of the Trombetas River in 1542.

Like the North American boreal forest, but for reasons unrelated to glaciers, the Amazon is a forest of water. It has 20 percent of the world's river water, which equals the volume of the Nile, Mississippi, and Yangtze combined. From December through April, high water turns 20 percent of the Amazon basin into a wetland. Every day its plants restock the sky with around 7 trillion gallons (27 trillion liters) of water vapor. The forest exhales a Lake Tahoe every five days. Many mornings you can see mist columns twisting up from the canopy as if from fairy campfires. This vapor coalesces into "flying rivers" that ride equatorial winds to the west, while at ground level the rivers flow down an extremely subtle slope eastward toward the Atlantic.

The previously accepted notion that vegetation is simply a consequence of climate with no reciprocal influence was upended by the Brazilian scientist Eneas Salati in the 1970s. He followed water across the basin by gathering rain samples in locations from the Atlantic to the Peruvian border and analyzing oxygen isotopes. Clouds drop the relatively heavy O-18 isotope first. If all the precipitation came directly from the ocean, rain far inland would be expected to have markedly less O-18 than that near the coast. In Salati's samples it didn't. He concluded that the forest was making its own weather, sending moisture from the original oceanic rain back up into the westward-traveling air mass to fall again five or six times. Salati showed that climate and forest are in conversation. When the Amazon megaforest's moisture finally finds the Andes, it rises, cools, and precipitates into the basin's most distant mountain tributaries. And there's still moisture left over, which curls north and south and irrigates ecosystems, farms, and cities in every country on the continent except Chile.

It's hard to know where to start with the Amazon's biological superlatives. The basin's bird species are estimated to number over 1,300. There are at least 3,000 species of fish, 370 reptiles, 420 amphibians, 430 mammals, and 40,000 plants. The round numbers are because no one really knows how many species are in the Amazon. Regions the size of European countries have yet to be surveyed. And the very idea of "species" is thrown into disarray when one looks—or listens—closely.

Take birds, for example. Mario Cohn-Haft speaks bird. The American ornithologist grew up in Western Massachusetts, rambling through

Channel-billed toucan in a flooded forest of the Mamirauá Sustainable Development Reserve, Brazil.

woods and bogs, always looking for something he'd never seen before. He ventured to the Amazon decades ago to study birds and still lives there, having found a lifetime supply of novelty. So far, he knows the calls or songs of around 3,000 Amazon birds—out of the 1,300 known to exist. That's right, Cohn-Haft knows more birds than there are.

He explains. After a boom in scientific bird description through the late 1800s and early 1900s, discoveries slowed to a trickle by 1950. Biologists thought they had found pretty much everything. "The 10,000 birds on the planet were a done deal," Cohn-Haft says. "But now, if you plot the curve, we're in a period of exponential growth, an explosion of discovery of new birds." Ornithologists formerly established the existence of separate species by looking at specimens in museum drawers, differentiating the mute creatures based on looks. Cohn-Haft immersed himself in the forest and found that different populations of birds—musician wrens, for example—sound different on one side of a river from look-alike populations on the other side. The definition of a species is the ability to produce fertile offspring. Cohn-Haft was observing birds that look the same but sing differently and don't breed, which has been verified by genetic testing.

"We keep discovering new birds and that's pretty neat, but we also are revising our notion of what species are," says Cohn-Haft. One species of musician wren became six. Whether they can't breed or prefer not to is unclear. Either way, it appears likely that the subtly diverging life-forms in countless river-bound wedges of Amazonia are in early stages of becoming wholly distinct organisms. "We're just constantly and radically revising our knowledge of every single thing we've ever found in the Amazon, including non-birds. Snakes and lizards and frogs and everything else," says Cohn-Haft. A new species of elaborately patterned harlequin toad, with skin toxin of some medical interest, was identified in 2020 near Manaus. The amphibian was established as a new species based on a combination of physical properties, its vocalizations, and a genetic comparison with close relatives for which it had been previously mistaken.

Several factors account for the biological richness of the Amazon. First are variations in climate and geology that give the rivers their black, blue, and white labels and determine how much mineral nutrition they carry. Also, Amazonian rivers are wide, which discourages the dispersal of poor swimmers, such as the Roosmalens' dwarf marmoset, a 6-ounce monkey first described in 1998 that is trapped in a triangle of jungle between the Madeira and Aripuanã Rivers. Even birds with respectable powers of flight will sometimes stop at the river's edge, refusing to traverse treeless space. Then there is a veritable species bonanza as the forest climbs the Andes. Microclimates are like steps in a grand staircase, each with its unique assemblage of creatures, trees, and microscopic beings. The tapir we saw swimming across a horseshoe of river in hot flat rainforest is just a few dozen miles from a spectacled bear—a creature she will never meet—snacking on orchid bulbs in a cool cloud forest up the Andean slope.

The Amazon can astonish the first time you walk into it—or after decades of visits. Over the years, Tom has spent countless days walking in the woods of his research camp in the jungle north of Manaus. He has seen just about everything there. One of the most dazzling, and also common, sights at the camp and throughout the Amazon is a blue morpho butterfly. The size of a flying postcard, it careers about the forest, reckless and beautiful, flashing iridescent royal blue wings as it goes. One evening in the fading light, as Tom prepared to receive a group of visitors, he glimpsed a morpho out of the corner of his eye. But something wasn't right. Its flight path was more inebriated than normal, the flier unusually pale and floppy. And big. So, man set off after insect through the forest. When it finally landed, Tom realized he was looking at the white witch, the largest moth in the Americas, and so rare that scientists have yet to identify its caterpillar.

In part, discoveries still abound in the Amazon because the region was mostly overlooked for the first four centuries of South America's colonial history. After several fruitless expeditions to find a fabled city of gold in the jungle during the 1500s, explorers and impresarios largely gave up and applied themselves to the exploitation of the continent's mountains and coastal regions. The notable exceptions were a trio of Victorian-era British naturalists, Alfred Russel Wallace, Henry Walter Bates, and Richard Spruce, who explored and collected specimens in the basin from the 1840s to the 1860s in a shared quest to understand the origin of species. Then, in the late 1800s, bikes became popular. They rolled on solid tires made serviceable by self-taught chemist Charles Goodyear's 1839 discovery that natural latex heated with sulfur produces rubber that is both elastic and stable. The process is called vulcanization after the Roman god of fire. Unprocessed rubber gets sticky when warmed and when cold becomes very hard. Tires that wouldn't stick to the ground in hot weather or crack in winter made cycling more fun and proved useful for another contrivance of the period: the automobile.

A sudden explosion of demand for rubber set off a mad rush to find latex-bearing *Hevea brasiliensis* trees. Some of the most productive concentrations of the trees were in the upper reaches of rivers in Colombia, Peru, Ecuador, Bolivia, and western Brazil, truly remote jungles that had been scarcely seen by European colonizers during their first 400 years in the New World. Indigenous peoples of the Amazon who had survived the plagues brought by earlier European explorers faced a brutal new invasion.

Collecting rubber requires visiting hundreds of scattered trees, scoring long diagonal grooves in the bark, collecting latex, carrying it back to camps, and combining it into massive superballs. These are then transported to the nearest navigable river. Rubber entrepreneurs sought workers able to survive in the forest and unable to negotiate good pay. Using violence, they enslaved some of the people they found there. Others were cajoled and encouraged into debt peonage. Some tribes, it should be said, managed a profitable trade with the outsiders, which gave them the upper hand in age-old wars with their neighbors. The rubber boom also attracted throngs of migrants into the Amazon.

In 1876, Englishman Henry Wickham took 70,000 rubber seeds from Brazil to the Kew Royal Botanic Gardens, where 2,700 of them successfully germinated. The seedlings journeyed onward to British colonies in Asia. In the Amazon a fungus had foiled attempts to grow *Hevea* plantations, but the fungus didn't exist in Malaysia and other tropical English territories. Planted rubber flourished there and, some four decades after the original seed export, had matured into trees that upended the economy of the Amazon. In 1910 the rubber price peaked, and in 1920 its bottom fell out due to the tsunami of cultivated supply coming out of Asia. The Amazon went quiet, economically speaking, for several decades.

Today, almost all Amazon deforestation emanates from roads. The Trans-Amazon and several other federal highways burrow through the forest, particularly in the south and east where the jungle meets the busier parts of Brazil. In the Andes, several roads wind from the highlands to rainforest cities, provoking extensive forest loss. Among the five megaforests, the Amazon is the only one where agricultural expansion, spurred by the roads, is the dominant factor in the fragmentation and loss of IFLs. Ranching and farming account for around 65 percent of the total. In the closest runner-up, tropical Africa, 23 percent of losses are due to farms and ranches.

Roads often begin as rough tracks bulldozed by loggers into a remote area. They selectively extract the most valuable trees. Settlers frequently buy fake titles made by operators called *grileiros* in Brazil. The name comes from the predigital method for falsifying land deeds: putting the papers in a box with defecating crickets (*grilos*) until the certificates looked antique. *Grileiros* now register bogus titles with several agencies at the same time such that one fraudulent claim validates another.

The homesteaders fell the remaining trees from several acres by the side of the muddy track. When the dry season comes, they burn the land and plant crops such as beans, corn, and a starchy tuber called manioc in the ashes. In a few years the farmers repeat the cutting and burning nearby, either to expand cultivation or to shift to fresher soil. They may diversify, planting some nutrient-poor but hardy African grass for a few cattle. A portion of properties is commonly kept forested to supply wood for fenceposts, buildings, and fuel. Some people may plant a few cocoa trees, which do well in the forest understory. When these settlers become numerous, they form a new community, often with a promised-land sort of name—New Hope, New Progress, New Life, Sweet Glory, or the name of a saint.

Soon, someone with deeper pockets may come along. Small farms with flagging productivity are bought up and consolidated. The remaining trees are cut, more grass is planted, and tropic-ready South Asian cows start to graze. At this point, at the urging of farmers both large and small, a municipal or state government may pave the road, which up until now has been a tire-sucking mud trap in the rainy season. There is often a final chapter: the half-ton cows wander around compacting and eroding the soil until it's barren. Or, if the land remains viable and is flat, a soybean farmer may buy the tract and cultivate the legume for export to Europe and Asia. Though there are many variations on this Amazon story, most involve some sequence of roads, logging, and farming.

During the first year of Jair Bolsonaro's presidency, 2019, Brazil's forest loss accelerated by 34 percent from the previous year. In 2020 it hit a twelve-year high. The new head of state spurred destruction by broadcasting his personal support for the activity and cutting enforcement budgets rather than by changing policies, which takes time and cooperation from Congress. The new numbers represented a steepening of the deforestation trend that had started under Bolsonaro's predecessors in the mid-2010s. Previous to that, Brazil had accomplished an astonishing 80 percent reduction in deforestation from 2000 to 2012. During that period, the country deployed a coordinated set of policies, had adequate budgets for enforcement, created protected areas, welcomed international funding, and largely put new rainforest roads on hold. Over the space of a decade, Brazil had shown the world, for the first time, that a country facing massive deforestation pressure could save its megaforest.

The Congo

Like the Amazon, the Congo owes its intact forest in no small part to a relative lack of roads. What roads it does have often lack bridges over the rainforest's copious broad rivers. Terrestrial travel, therefore, takes a certain amount of determination, patience, and surrender to the whims of the ferryman. That's why John and a group of companions spent an hour in the fading golden light of a November 2019 afternoon racing along dirt roads through the jungle at white-knuckle speeds.

The ferry that carries cars, trucks, and people across the Sangha River runs only in the daytime. We were on our way back from a weeklong visit to one of the Congo's most intact, wildlife-rich forest redoubts, called the Goualougo Triangle, which is separated by the Sangha from the country's paved roads, towns, and "civilization" in general. When we reached the crossing, we maneuvered around several trucks loaded with lumber, manioc, and charcoal. Across the water, lights were winking on in the small city of Ouesso. Our driver dismounted to chat with someone called "Manager" on the concrete ramp. Calls were made with people on the other side to see if the ferry could be prevailed upon to make another run; we had no apparent sleeping options on this side of the river other than crammed in our vehicle. To the east, a landscape of forests and wetlands—with scattered villages—stretched for hundreds of miles. Just as the sky ignited in a truly lovely weave of salmon, red, and gray, we received the news that the ferry was done for the day.

There was an alternative, it turned out, a boat called a pirogue that might still cross that evening. To seek the pirogue, we drove a short way upstream to a village called Djaka. Manager (this turned out to be his name, not a title) guided us. Soon we were unloading our bags in the dust, amusing the residents of Djaka as they relaxed by front doors in the twilight. Kids bathed in the shallows around two massive trees growing straight out of the water.

As we waited, a pickup truck dropped off a full load of passengers, who streamed down to the waterside with packages. A taxi delivered three beautifully dressed women with various children who all walked down to wait for the pirogue. A scholarly looking young man in wire-rimmed glasses appeared, carrying a goat, and proceeded down the hill. We asked Manager if perhaps we shouldn't heave our suitcases down the bank and stake out an advantageous spot in the mud to assure places on the pirogue. He brushed aside the suggestion at first and then, as more prospective passengers amassed, agreed, and we moved our hillock of belongings closer to the water.

At length, we spied the pirogue's bow light as it nosed quietly past the grand trees and approached the shore. It looked skinny. Like a canoe. The idea of all the well-dressed women and the dozens of other people milling about in the dark, as well as our party of four Americans and four Congolese conservationists, and the scholar and his goat, and Manager, all boarding this slender craft was preposterous. At Manager's signal we hoisted our luggage and made our move toward the pirogue's bow, which was now beached in the mud. There was a great deal of shouting in French

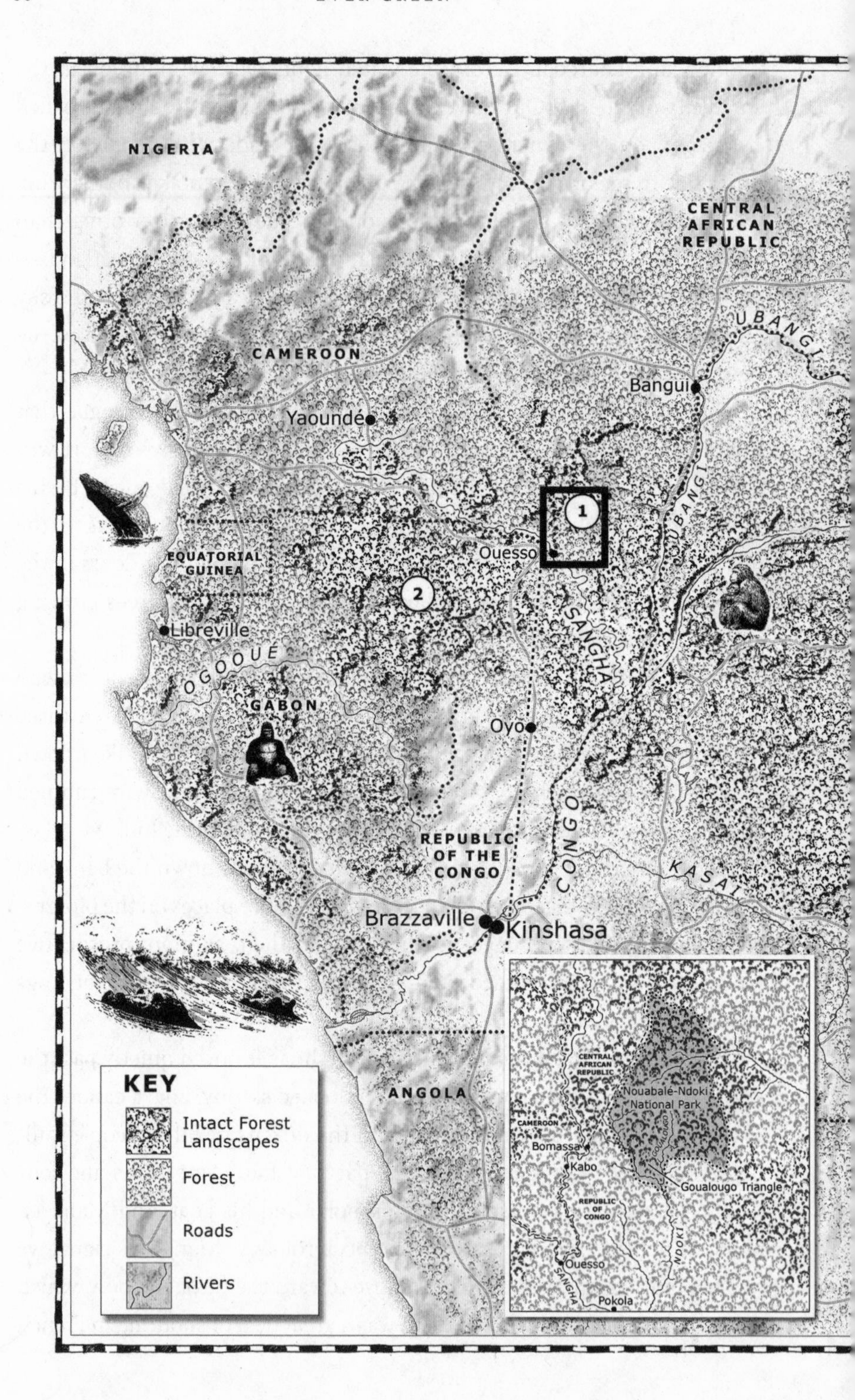
NIGERIA
CENTRAL AFRICAN REPUBLIC
UBANGI
CAMEROON
Bangui
Yaoundé
1
EQUATORIAL GUINEA
Ouesso
2
SANGHA
Libreville
OGOOUÉ
GABON
Oyo
REPUBLIC OF THE CONGO
CONGO
KASAI
Brazzaville
Kinshasa
ANGOLA
KEY
Intact Forest Landscapes
Forest
Roads
Rivers
CENTRAL AFRICAN REPUBLIC
Nouabalé-Ndoki National Park
CAMEROON
Bomassa
Kabo
Goualougo Triangle
REPUBLIC OF CONGO
NDOKI
Ouesso
SANGHA
Pokola

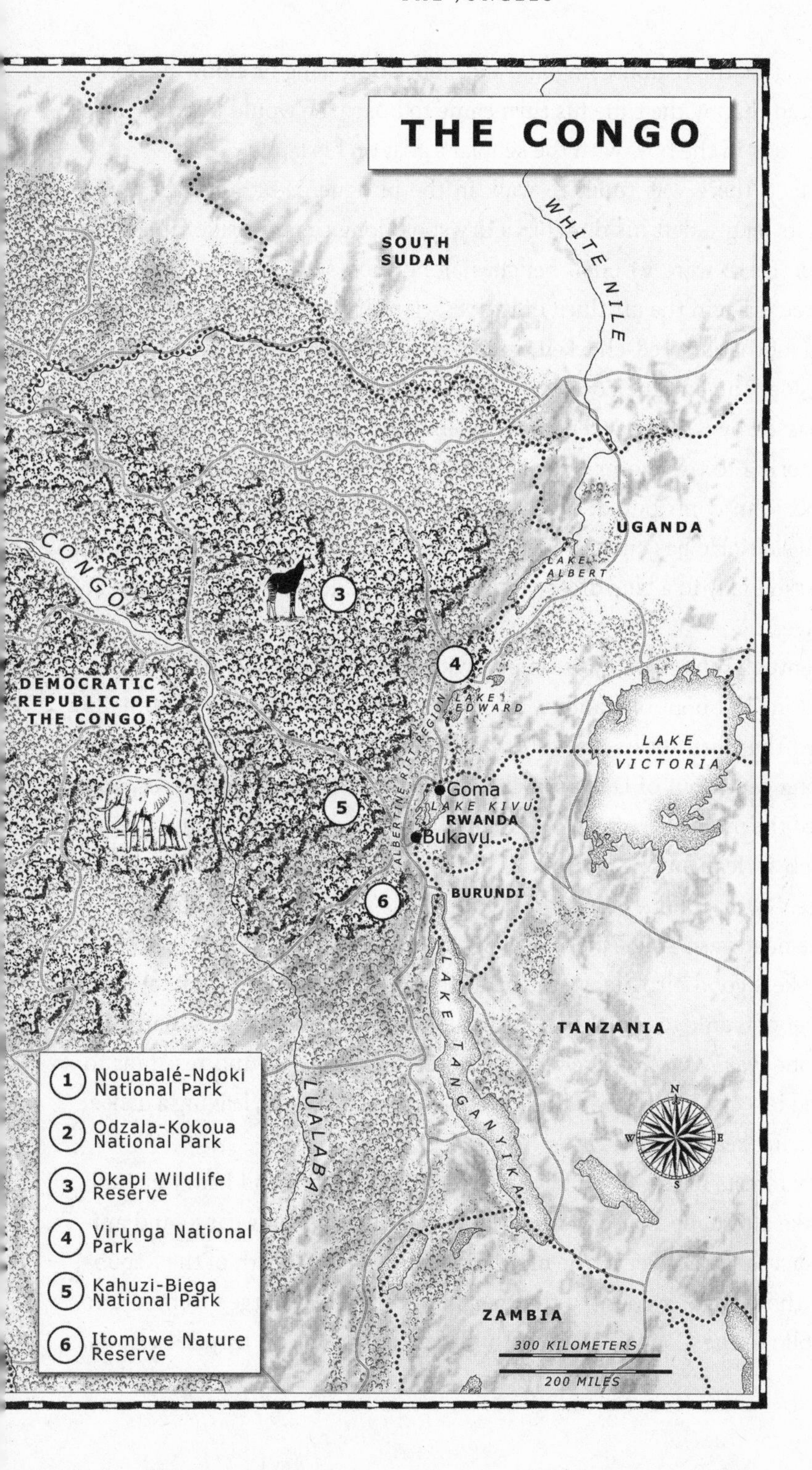
THE CONGO
SOUTH SUDAN
WHITE NILE
UGANDA
LAKE ALBERT
CONGO
DEMOCRATIC REPUBLIC OF THE CONGO
LAKE EDWARD
ALBERTINE RIFT REGION
LAKE VICTORIA
Goma
LAKE KIVU
RWANDA
Bukavu
BURUNDI
TANZANIA
LAKE TANGANYIKA
LUALABA
ZAMBIA
300 KILOMETERS
200 MILES
N
W
E
S
1 Nouabalé-Ndoki National Park
2 Odzala-Kokoua National Park
3 Okapi Wildlife Reserve
4 Virunga National Park
5 Kahuzi-Biega National Park
6 Itombwe Nature Reserve

and Lingala, the regional lingua franca, but no shoving or elbowing. John expected that by the time his turn came to board, he would perhaps luck into a nook in the bow with the scholar's goat on his lap.

In fact, there was room to walk in the pirogue. He ambled partway down its length, left his duffel in a dry spot along the port side, and continued. There were wooden benches along each side, and he continued to a free place in the aft third of the vessel, near the matrons. He sat by a young mother in a red-checked dress and matching headgear with her son alongside. The pirogue continued absorbing the crowd from the bank. A boy passed by with a saucepan of gasoline and handed it to a committee ministering to the motor by flashlight. Shouting continued. The women around John contributed, commanding people to move back, stand up, switch sides. He has enough French to guess what to do. He stood up. An impressive lady in a blue dress ordered him to sit down. People shouted at Manager.

Eventually the beach was empty, with the pirogue's bow firmly stuck on it. One last bout of standing up, sitting down, and shifting of humans toward the stern set us free. The motor purred as the pilot maneuvered the long canoe out of Djaka's swimming hole, past the ancient trees, and toward the main current of the Sangha. The stars and night air and maybe a touch of fear among the nonswimmers charmed the passengers into silence. Several phone screens dotted the boat's length with pools of light.

The boat was a single piece of wood around 65 feet long and 5 feet wide. To make a good pirogue, you need a tree much longer than that, at least 120 feet of trunk, to avoid cracked and twisted bits, and a 6-foot diameter. This one was *Nauclea diderrichii*, whose wood is heavy and resistant to rot and insects. The tree was hollowed down to a flat floor, leaving a dense, reassuring keel below.

Few forests in the world have trees big and straight and heavy enough to make a pirogue that can carry six dozen humans, plus cargo and animals, across a big river in the night. The woods in the north of the Republic of the Congo, a country that neighbors the much larger Democratic Republic of the Congo, is one such place. Of the world's tropical forests, the Congo has the biggest trees by volume. As soon as you walk into the woods, you're among giants. Finlike buttresses flare at the bases of some. Others emerge fat and round straight from the earth, as if their boles extend far underground before eventually giving way to roots. For one accustomed to the Amazon's weave of palms and vines, the Congo forest is immoderate, grand, and vertical.

This is the planet's second-biggest rainforest. The Congo basin contains half a billion acres of moist tropical forest and another 250 million acres of fringing secondary forest and dry woodlands on the adjacent plateaus. Around 60 percent of the rainforest is in the Democratic Republic of the Congo, around 10 percent each in the Republic of the Congo, Gabon, and Cameroon, and minor shares in the Central African Republic and Equatorial Guinea. The Congo River is born in the mountains in East Africa and describes a nearly 3,000-mile parabola, running northwest before bending southwest to a set of cataracts just inland of its mouth. Biodiversity is highest in the western and eastern extremes of the Congo: Gabonese Atlantic coastal forests are the wettest and richest in species among the basin's lowlands. Elephants, leopards, gorillas, and forest buffalo walk right out of the forest and stroll on Atlantic Ocean beaches. Hippos bodysurf.

In the east, the Albertine Rift foothill and montane forests are diverse thanks in part to the variety in terrain, soils, and altitude. The area around this geologic fissure is a species treasure chest. It is home to over half of Africa's birds, 40 percent of its mammals, and a stunning diversity of butterflies. Emma Stokes, head of Central Africa programs for the Wildlife Conservation Society, says, "It's this crazy zone of super-high biodiversity in the eastern Congo. You have okapi, you have mountain gorillas, you have Grauer's gorillas. Who knows what birds and herbs and everything else! Every park you go to has something unique and *big* and different in it. Not just like a unique bird, but a unique gorilla."

There's also more pressure there. Eastern Congo has three things the western reaches of the forest lack: massive mineral wealth, fertile soils, and lots of people. In the eastern DRC provinces of Ituri, North Kivu, and

South Kivu, populations range from 173 to 251 people per square mile. Gabon, on the Congo basin's western end, has 22 humans for every square mile, a third of whom live in the capital city, Libreville.

And the eastern region lacks one thing that is present in at least basic form in the western Congo: a government. In the west, the authorities in Gabon, the Republic of the Congo, and Cameroon control their territories. In the eastern DRC, far from the capital Kinshasa, freelancing Congolese soldiers, various militias, and forces from neighboring Uganda and Rwanda exercise a high degree of de facto rule. Eastern forest losses reflect these challenging conditions. The DRC lost 4.2 percent of its primary forest between 2001 and 2018. The western Congo countries of Republic of the Congo and Gabon lost only 1.4 percent and 1 percent, respectively.

The western Congo has Africa's biggest populations of forest elephants and great apes, along with various other primates, forest antelopes (called duikers), and a safety-vest-orange, floppy-eared wild pig. The western Congo also has hundreds of thousands of acres of monodominant bemba (*Gilbertiodendron dewevrei*) forests. As a rule, tropical forests mix dozens and sometimes even hundreds of tree species in a single acre. Here, the often-colossal bembas account for 80 percent of stems in some wet, infertile spots. The tree deploys a couple of tricks to pull this off. The first is called "light plasticity," which just means that its seedlings do well either in shade or sunlight. The second is a relationship with underground fungi that permits the bemba to build itself out of mineral-poor soil. Once a bemba patch gets going, it tends to stick: mud cores have shown bemba pollen dating back 2,700 years in a given spot, while nearby mixed forests show evidence of frequent species turnover in response to climate shifts.

The basin's western lowlands are distinguished by another remarkable feature only "discovered" in 2012 even though it's bigger than England: the world's largest tropical peatland. In swamp forests shared by the Republic of the Congo and the DRC, this heretofore hidden system has trapped around 30 billion tons of carbon, equal to around twenty years of US fossil fuel emissions.

A hundred and sixty years ago, Europeans knew almost nothing about the Congo. Thanks to thick forests and the impassable waterfalls near the river's mouth, the basin was a nonmetaphorical blank spot on the map. Pygmies,* the people who have been living in the Congo the longest, were no more real in the European imagination than wood sprites. Outsiders suspected that the continent's two great rivers, the Congo and the Nile, had a common source. Henry Morton Stanley was the first White man to traverse the basin. On his first expedition, between 1874 and 1877, the explorer established that the Congo River had nothing to do with the Nile. Shortly thereafter, Stanley entered the service of King Leopold II of Belgium and claimed for his employer 60 percent of the river basin in a colony called the Congo Free State. Joseph Conrad, who captained steamships on the river for three years in the 1890s, distilled the brutality of Leopold's reign into two words pronounced by the *Heart of Darkness* hero, Captain Marlow: "imbecile rapacity." The fictional skipper describes the real European practice of standing on the deck of a ship and shooting local people for amusement.

Slavery had been abolished by all European nations by the time King Leopold II acquired his personal colony under the pretext of a nonprofit mission to bring God and progress to the peoples of the Congo. Leopold enslaved them instead. His agents would kill and maim people until the rest agreed to collect ivory, and later rubber, for free. Then they would kill and maim workers for not collecting enough ivory and rubber. Around 10 million Congolese, half the population, perished in thirty years. The rubber genocide was an African mirror of the boom taking place in the Amazon.

Eventually, Edmund Morel, a clerk for the Liverpool firm with the monopoly on Congo shipping, noted that vessels returned to the colony with guns and ammunition for Leopold's agents, not goods the workers

* "Pygmy" is an umbrella term used to refer to various peoples of generally small physical stature whose lineages in the Congo extend back tens of thousands of years. The name is commonly used in both academic and informal settings without pejorative connotations.

could plausibly buy with their wages. Morel embarked on a long, ultimately successful crusade against the Belgian king's slave system. Leopold became the object of outrage and was shoved out of the business—with a handsome severance from the Belgian parliament. But his murderous system was adopted by the French in the present-day Republic of the Congo, Gabon, and Central African Republic and by the Germans in the area that is now Cameroon. Historians believe a similar proportion of the human inhabitants of those territories died as a result. Further, after Leopold was gone, the Belgian government took over the business and continued to run it with unwilling workers, coerced into service by a head tax they couldn't possibly pay except by working in the government's enterprises.

Our guide for the trip to the Goualougo Triangle was Dave Morgan, who has been researching chimps and gorillas there since the late 1990s. Morgan got his start in zoos. While attending college at Western North Carolina University, he worked at a roadside zoo, a job that involved, among other tasks, rebuffing religious zealots who wanted to pick up the venomous snakes. After college he went home to Florida and worked at Busch Gardens, a zoo and theme park owned by the St. Louis brewing family. Old-timers told him stories about drinking free Budweiser with chimps in the breakroom. Though Morgan would never do that, one can imagine the former college linebacker fitting in. His English, French, Lingala, and Latin (for plant names) all come out in an accent he describes as a mix of "Redneck Riviera and Appalachia."

In 1996, Morgan got his chance to work in the wild, joining a biologist named Mike Fay, who was managing the Nouabalé-Ndoki National Park in northern Republic of the Congo. Morgan initially worked near the Goualougo Triangle, but not in it. That's because Fay had declared the area off-limits even to research. When Fay had wandered out there in 1991, he found "naive" chimps and gorillas, a primatology term for apes that have no previous experience with humans. They didn't run away from Fay or try to run him off. Fay wanted the Goualougo Triangle to stay innocent.

But in 1999 it became clear that a logging company, Congolaise

Industrielle des Bois (CIB), was going to start logging near, and perhaps even in, the Goualougo Triangle. The biologists were faced with the reality that the forest jewel would be logged if they didn't occupy it. As part of a pragmatic deal struck with the logging firm, Morgan was tasked with setting up the Goualougo research project.

To get there we drove for two days from the capital Brazzaville, managed to catch the same ferry across the Sangha River that we would narrowly miss on the return journey, and drove 3 more hours on orange dirt roads that narrowed until the bushes slapped the Land Cruiser's mirrors on both sides. At the road's end, we climbed into small pirogues with local Pygmy paddlers and wound our way up the slow and swampy Ndoki River and then the smaller Mbeli, under tree arches in a channel that smelled like a garden and dwindled in spots nearly to the width of our canoes. After an hour we disembarked and started walking on trails made by elephants.

These smooth paths, perfectly cupped into the forest floor, bend elegantly this way and that, converging at four- and six-way crossroads. They give the woods a feeling of festive occupation and make hiking easy. The only obstacles of note are the trail makers' dung. These auburn splats look like shredded redwood bark mulch and smell ripe but not offensive. They're the size of garbage can lids, give or take, often sprouting tiny trees and mushrooms. Occasionally, the elephants drop more cohesive coffee-can-sized pellets.

Following one of these trails, you can almost imagine an elephant's day, visiting one tree after another. Furry octagonal *Duboscia macrocarpa* fruits are chewed and scattered. The next tree announces itself with a burnt-sugar-and-cherries aroma. Long reddish-black pods of *Tetrapleura tetraptera* are strewn half-chewed on the trail and in the bushes all around its base. Elephants cut the trail to this tree, but gorillas love the seeds, too, inserting a canine and running it the length of the capsule. People gather them for food, medicine, and for the pods' aromatic properties. The tree can do just about anything—reduce inflammation, convulsions, blood pressure; function as birth control; and deter the schistosomiasis parasite, which causes digestive misery and even death. The sticky pods might

Pygmy tracker paddling the Ndoki River.

simply taste good, but gorillas and elephants, like many other species, are known to use plants medicinally. Further along, we come upon a "scratching tree," deformed and stuccoed with red dirt that elephants throw on their backs and rub off on selected trunks.

Forest elephants are a separate species from those that roam the savannas. *Loxodonta cyclotis* is adapted to moving through the woods, with smaller bodies and straighter tusks than their nonforest relatives (*L. africana*). What really hits you after walking on their trails for a while is a realization that they engineer the entire forest for the mobility of other species, including us. The elephants create a sort of faunal ductwork. We found leopard and forest buffalo droppings on the trail and smelled the cotton-candy-scented pheromone emitted by red duikers. Chimps, gorillas, bongos, various other antelopes, and wild pigs all use the elephant paths to get around.

Six hours' march brought us to a place where the trail disappeared underwater. It was a river inside of the forest, a swamp with a current, and our final obstacle before reaching camp. We took off shoes and waded into the Zoran River. The swamp slowed the wheels of the mind. The canopy lowered to a more intimate height. Conversations paused. The water cooled our ankles, then knees, then hips as we quietly immersed and moved in a line, feeling our way tentatively in the opaque coppery water. The bed was smooth and sandy in places, padded elsewhere by mats of leaves. Unseen roots and logs were navigated via a friendly system of hand-signaling to the next person in line. Before the wade, Dave Morgan had recounted crossing a similar river in Goualougo Triangle one day when something in the dark water suddenly bolted through his legs. When it surfaced, he caught a glimpse of a water chevrotain, a striped and spotted dog-sized deer, known to hide from predators in the water.

After 20 minutes, we walked out of the swamp, a little sorry it was over. We laced up our shoes and after a few more minutes were ducking under an elephant fence—wires festooned with empty sardine cans—that forms the perimeter around the Goualougo Triangle research camp. The camp's bustling population included two dozen Pygmy trackers—several at a spirited card game—four or five studious Congolese research assistants, a cook, and camp manager.

For most people, simply being near a gorilla or chimp with no fence involved is a mind-blowing experience. Morgan, on the other hand, has been around the apes for over two decades. He's had every kind of encounter, including once when he "got slapped around" and sustained a serious gorilla bite to the shoulder. His peak moments are when he sees the primates do something unusual. Novel ape behavior expands the human mind and often shrinks the domain of what people have doggedly attempted to claim as uniquely human, such as humor, mourning, cooperation, planning, tool use, and other special things we do. In fact, apes do all these things. They tickle, prank, and laugh. In the first decade of their research at Goualougo, Morgan and his wife, Crickette Sanz, documented twenty-two different instances of tool use among chimps, including tools for feeding, self-care, first aid, comfort, and play.

Over dinner our first night in camp, Morgan explained that we might witness a behavior few humans have ever beheld: ape co-feeding, which is when two ape species eat together. Co-feeding involving gorillas and chimps was not known to occur a couple decades ago and is being documented exclusively here in the Goualougo Triangle, the only place in Africa with both chimps and gorillas that have been habituated to the presence of researchers. For days, Morgan's team has been monitoring a massive *Ficus recurvata*, a fig tree about 30 minutes' walk from camp. This tree is one of the many strangler figs native to the tropics. They grow from seeds that fall, or are defecated, into a high nook of another species. The fig seed germinates and starts life with excellent access to light, water, and just enough trapped bits of soil and decaying leaves. It sends root tendrils to the forest floor, 50 feet or more below. These roots anchor, thicken, and wrap the host tree, which perishes and leaves a hollow center. By that time the weave of *Ficus* wood is sturdy enough to hold up its own spreading canopy.

The fig tree near the Goualougo camp produces golf-ball-sized fruits. They smell like a smoothie made of green mangoes, cardboard, and cut grass. A botanical survey in the area found only one of these trees in every 1,500 acres. They fruit once every thirteen months, and the chimps have the locations and dates marked on their mental calendars. In the days preceding our visit, the research teams observed a female chimp coming by several times to test the figs for ripeness, like someone squeezing supermarket peaches.

"It is goin' off!" Morgan announced around the campfire on the eve of our first outing to see the animals.

We set off before dawn the next morning, falling into a line of researchers led by Pygmy trackers. When we arrived, the fig tree held two gorillas and four chimps, plus some vivid green pigeons and noisy birds called giant plantain eaters. Morgan was exultant. "It's happenin,' people!" Research assistants entered behavioral data on tablets. This work

Chimps and gorillas, obscured by foliage, are feeding together in this *Ficus recurvata*.

is done by teams of Congolese biology students and Pygmies from the vicinity of the Nouabalé-Ndoki National Park. For the novice observer it was not immediately clear which species was which. They were the same dark color, the chimps were bigger than expected, and the gorillas aloft were young males. Botanist David Koni pointed out who was who. Koni also inserted twists of wild ginger into John's ears to keep swarms of salt-loving sweat bees out. Or to make him look funny. It accomplished both.

Co-feeding is not a lovefest. There is an obvious power imbalance. The chimps are canny, fast, territorial, and move in large communities. The group we met is called the Moto community, after the Lingala word that means both "fire" and "man." It has fifty-five members, around half of whom were present. Gorillas live in smaller family groups, like the company of five we observed, led by a silverback called Loya.* The chimps seemed, at best, to be tolerating the gorillas. The gorillas in the trees, Loya's sons, Mojai and Kao, ate with one eye on the fruit and the other on the chimps. At one point Mojai got into a pickle. Hugging a fat smooth branch, he slid down to the place where the fig tree first bifurcated, far above the forest floor. Soon there were chimps directly above him on both of the main branches, and the central trunk was too big for him to embrace for further sliding. He nervously looked for options and tested his (insufficient) reach around the tree's circumference. Eventually, the chimp on one of the exit branches moved off and Mojai scrambled up and into a neighboring tree. The rest of the gorilla family settled for fruits scattered on the ground.

This towering fig specimen draws the two ape species into a temporary bubble of intimate, if uneasy, coexistence. While in proximity, they benefit from each other's alarm calls. The younger individuals will engage in cross-species play and even sexual exploration. There is chest-beating and sometimes violence. Gorillas are known to benefit from chimps' awareness of ripening fruits and follow them to the trees. And the researchers

* All the studied apes have names, chosen by the trackers and research assistants based on an animal's character and/or as an homage to certain admired people. Loya is named after the Goualougo project's lead tracker, Loya Gaston.

are watching for any sign that the gorillas pick up on the chimps' tool use. Nothing conclusive so far. Morgan also points out that what we watched was a dynamic that could very well have been part of the Pleistocene human experience, during which *Homo sapiens* had human contemporaries such as Denisovans and Neanderthals. "Bones can only tell us so much and seeing real behavior in the wild is instructive. All we have to go by at this point is sympatric chimpanzees and gorillas." The term *sympatric* describes species that inhabit overlapping areas.

We moved away from the ficus with David Koni to look for Loya. The silverback was asleep, sprawled on the forest floor. Koni is a plant savant, able to identify almost anything in the forest with his eyes shut. He whispered that when Loya woke up, the ape could reach out in any direction and find something edible, from the gorilla point of view. Like cows, gorillas ferment their food during digestion—in the apes' case within oversized colons. They can eat just about anything green. As if on cue, Loya stirred, rolled, reached out a huge arm, and stripped leaves from a nearby bush. He munched, stood, and then ambled along an elephant trail, back to the co-feeding tree.

The most shocking thing Dave Morgan has to say about deforestation in the Congo is that he's never seen any. The idea that a tropical conservationist with twenty-five years of experience is still waiting to see his first smoking clear-cut is unfathomable. But as we drove here, we didn't see any either.

Morgan explained, first of all, that few people live in the environs of the Nouabalé-Ndoki National Park. Second, the local population is largely comprised of Indigenous Aka and Baka Pygmies who traditionally have lived in and from the forest. They are among the most numerous of the Congo's original forest peoples, which include over a dozen distinct ethnolinguistic groups. Genetics suggest that Pygmies have a common ancestry that diverged from other human groups around 60,000 years ago, splintered into separate regional populations 30,000 years ago, and, around 3,000 years before the present, interbred to a limited extent with taller migrants arriving from the north. Ever since that contact, Pygmies and the immigrant farmers have maintained strong associations and separate cultures. Pygmies have been the forest guides, hunters, foragers, spiritual intermediaries, and, in some instances, underclass for the farming cultures to which they are linked.

The farmers in the park's vicinity pertain to the massive Bantu linguistic group, whose ancestors came from West Africa. They grow manioc and other foods on a small scale and get the rest of their sustenance, often through barter with the Pygmies, from the forest and rivers. Global commodities such as beef, soybeans, palm oil, and cacao that have caused havoc in other tropical forests have, as yet, no firm foothold in the Congo.

Logging, on the other hand, is a pervasive threat across the six countries of the Congo rainforest. In 2002, the DRC declared an industrial logging moratorium to curb environmental mayhem and choke off a funding stream for armed conflict. But felling flourished in the basin's largest country through the 2010s at eight times the officially acknowledged pace, with 90 percent of all timber harvest violating the country's forestry laws. In 2018, the DRC began reinstating industrial concessions, starting with licenses held by two Chinese firms for forests that grow in the sensitive central Congo peatlands.

Logging is often highly selective, with as little as a tree taken per acre. Once loggers are finished, some of their roads are reclaimed by the forest within a decade—but not, in many instances, before hunters avail themselves of easier access to net, trap, or shoot forest antelopes, monkeys, apes, elephants, pigs, armadillos, pangolins, and rodents. Selective logging damages only around five percent of the forest, but intensifies hunting on nearly 30 percent.

In 1994, Karl Amman, a Swiss activist and photographer, and Gary Richardson, of the UK-based World Society for the Protection of Animals, stumbled on evidence that the logging firm Congolaise Industrielle des Bois (CIB) was complicit in commercial hunting of great apes in the Republic of the Congo. CIB, whose logging concessions border the Nouabalé-Ndoki National Park on three sides, protested innocence. But the company had, in fact, brought thousands of people into the forest, cut access roads, and encouraged hunting as a practical matter. There was no other animal protein available for the firm's workers.

CIB subsequently established a partnership with the Wildlife Conservation Society to spare wildlife in its concessions around Nouabalé-Ndoki. It earned a green certification seal from the Forest Stewardship Council (FSC), the leading international standards body for forestry. CIB is now doing better than the handful of other timber concession holders in the Republic of the Congo. It helps pay for anti-poaching patrols and, in consultation with local communities, designates legal hunting zones for nonendangered game animals on which Aka and other locals rely. CIB shares forest data and plans with the Wildlife Conservation Society, which jointly manages the national park with the government.

Morgan and colleagues recently published a paper showing minimal impact on chimp and gorilla populations in CIB's logging leases after a single round of logging sapele—which accounts for 70 percent of felled trunks. Both ape species remained abundant. WCS's Emma Stokes avers that the certified concessions, with their anti-poaching patrols, are enormously important for elephants. "And for bongos, which you don't find inside the park. Bongos love those clearings in the secondary forest and, as long as they're not hunted, they can thrive in concessions."

But the verdict on sustainable Congo logging, based on the view from space, is less encouraging. "FSC is not the solution, especially for Africa. It's just not," says Peter Potapov, referring to the certification program. His 2017 *Science Advances* paper found that FSC-certified logging in Gabon, Cameroon, and the Republic of the Congo has fragmented IFLs equally or more than forestry done without the green seal. In 2014, the FSC passed a motion to avert fragmentation of intact forest landscapes in certified operations. In the Congo this rule is ignored. CIB, the largest certified operator in the Republic of the Congo, is building new roads into areas that until very recently qualified as IFLs. Greenpeace International, a founding FSC member, pulled out of the alliance in 2018 over this issue.

Roads are the fly in the ointment of "sustainable" logging. In part, the Congo jungles have been protected "passively" up until recently by lack of access and benign neglect. They are remote and sometimes dangerous,

Silverback Metetele feeding from a giant *Duboscia macrocarpa* near the Goualougo project's Mondika camp.

particularly in the eastern DRC, so their conservation hasn't required active regulation and patrols. But human populations throughout the basin are growing at some of the highest rates in the world. And countries have a ready investor in infrastructure that will induce movement to hard-to-reach spots. In a single day's drive in the Republic of the Congo, we saw two airports, a stadium that looks like a hamburger, a convention center, a gleaming and very large university campus, and hundreds of miles of road, all built by firms from China.

New Guinea

In contrast to shrinking ecosystems worldwide, which we often think of as habitat islands, the Amazon and Congo megaforests strike many as seas of forest in the middle of the planet's two great southern continents. The smallest of the five megaforests, New Guinea is in a category of its own: a sea of forest covering a real island. To locate it, find Australia on the map, look northeast, and follow the finger of land called Cape York Peninsula. It points at New Guinea across the shallow Torres Strait. The planet's second-biggest island totals around 200 million acres, which equals two Californias, or slightly less than half a Congo. Remarkably, though the island megaforest is easy to access on all sides, it is still relatively intact five centuries after a Portuguese mariner was the first European to spot it, in 1511. Other tropical islands of comparable dimensions, such as Borneo, Sumatra, and Madagascar, all have lost most of their trees.

The eastern half of the island is Papua New Guinea, which gained independence from Australia in 1975 and is often referred to by its acronym, PNG. Earlier colonial masters were Britain and Germany. The island's west is ruled by Indonesia, a nation of 17,000 islands. Newly independent from the Netherlands in the 1950s, Indonesia asserted a claim to Dutch New Guinea. In the United Nations, where the debate played out, the Netherlands argued against it, pointing out that the Papuan people were culturally distinct and should rule themselves. The dispute was settled in the early 1960s when the United States, making a Cold War geopolitical calculation, opted to quietly support Indonesia. The half-island came

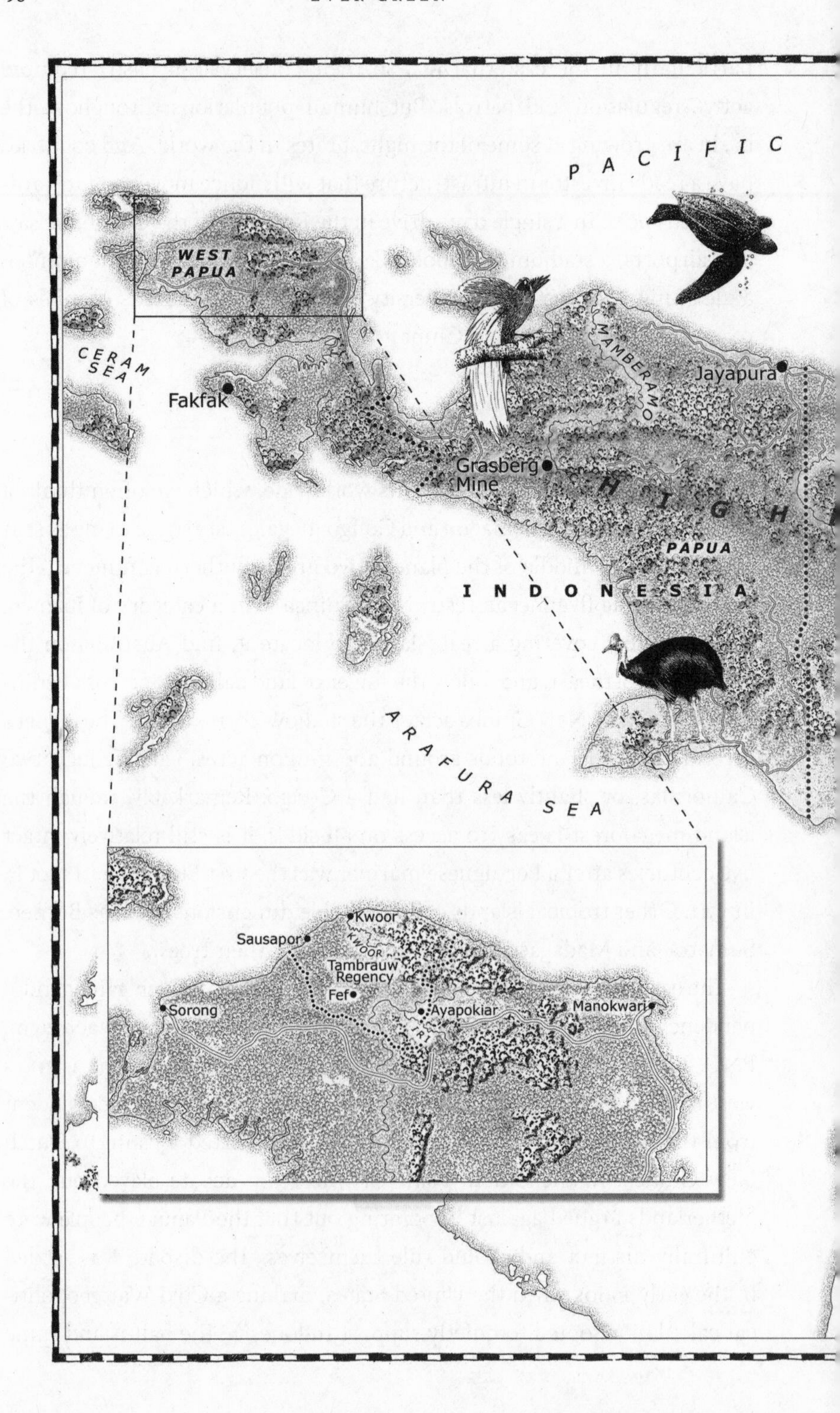
PACIFIC
WEST PAPUA
CERAM SEA
Fakfak
MAMBERAMO
Jayapura
Grasberg Mine
HIGH
PAPUA
INDONESIA
ARAFURA SEA
Kwoor
Sausapor
KWOOR
Tambrauw Regency
Fef
Sorong
Ayapokiar
Manokwari

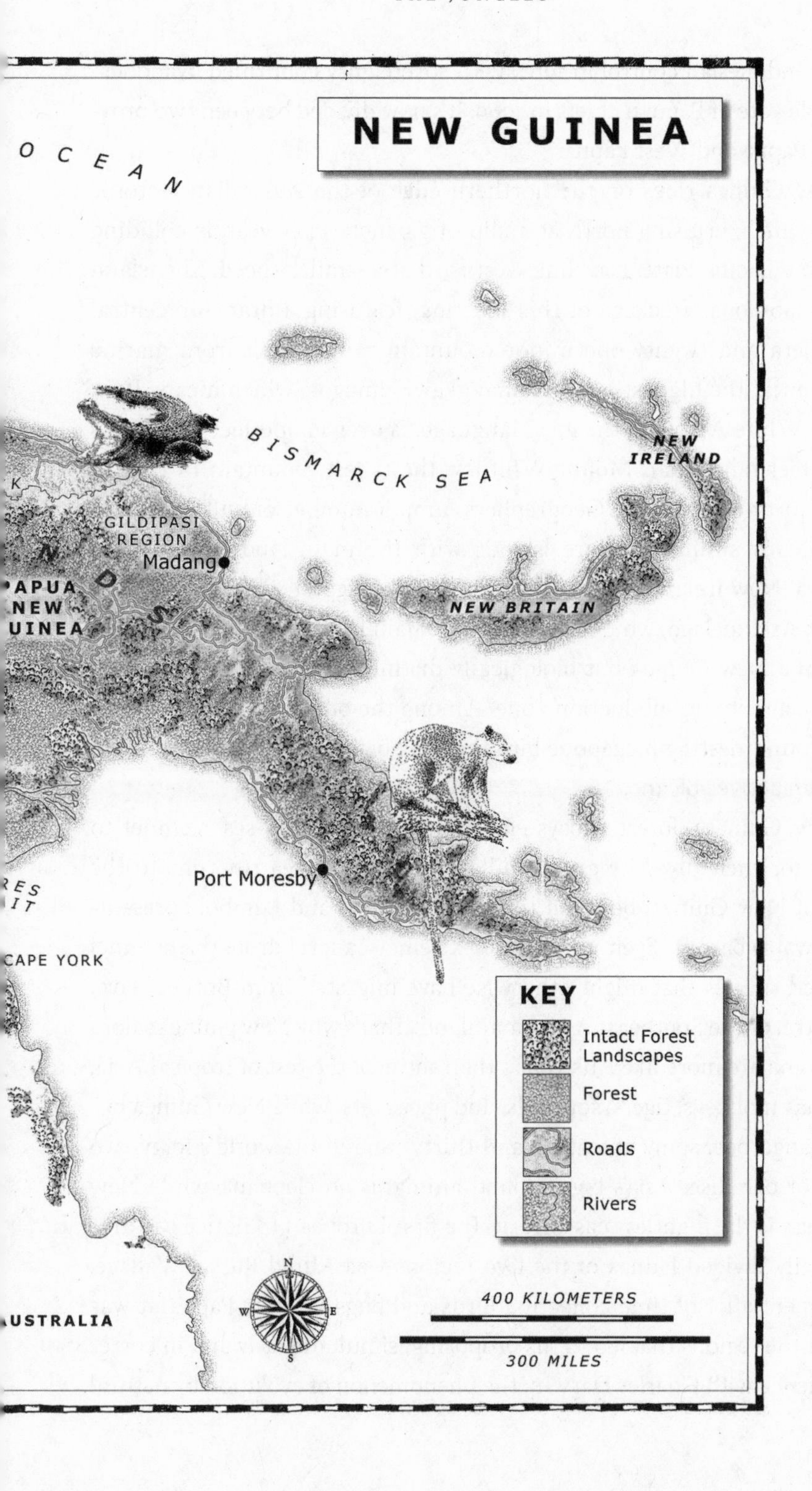
NEW GUINEA
OCEAN
BISMARCK SEA
NEW IRELAND
GILDIPASI REGION
Madang
NEW BRITAIN
Port Moresby
CAPE YORK
KEY
Intact Forest Landscapes
Forest
Roads
Rivers
N
W
E
S
400 KILOMETERS
300 MILES

under Indonesian control in 1963, with sovereignty confirmed by a questionable vote of Papuan chiefs in 1969. It's now divided between two provinces, Papua and West Papua.

New Guinea rides on the northern edge of the Australian tectonic plate, which, cruising north at a clip of 2.5 inches per year, is colliding with the Pacific Plate, traveling westward at a similar speed. The island is the fabulous wreckage of this meeting, featuring a dramatic central cordillera and twenty-one minor mountain ranges built from marine sediments. The highest peak, Nemangkawi Ninggok, which means Peak of the White Arrow in the local language, is over 16,000 feet tall. That's 1,500 feet taller than Mount Whitney, the tallest mountain in the US lower forty-eight states. Geographers lump a number of culturally and biologically similar offshore islands with the main landmass of New Guinea. New Ireland and New Britain are the biggest islands in the Bismarck Archipelago, which is close to the main island and politically part of Papua New Guinea but biologically distinct thanks to a deep ocean trench, a tectonic subduction zone. Among the oddities on New Britain is a ground-nesting megapode bird that incubates its eggs using the heat from an active volcano.

New Guinea's forest ecosystems have another deep-sea channel to thank for their diversity and peculiarity. The undersea rift runs to the west of New Guinea, between the islands of Bali and Lombok, presenting a water barrier. Even in times of extreme sea-level drop, the channel stopped species that might otherwise have migrated from Borneo, Java, Sumatra, or the Southeast Asian mainland. That's why New Guinea's flora and fauna are more like Australia's than those of the rest of tropical Asia. Asia has monkeys, tigers, squirrels, and pheasants, while New Guinea has tree kangaroos, spiny anteaters, and thirty-nine of the world's forty-two birds of paradise. Asia's biggest land animal is an elephant, while New Guinea's is the flightless cassowary. The first European to notice the dramatically divided faunas of the two regions was Alfred Russel Wallace, who spent a lot of time collecting birds and bugs in West Papua. It was one of the regions that led to his proposing, simultaneously and in correspondence with Charles Darwin, the phenomenon of evolution by natural selection in 1858. The pieces all fell into place for Wallace while in the grip of a malarial fever on an island just west of New Guinea.

The island megaforest's physical diversity has given rise to at least seven distinct forest types, as well as marshes, a savanna, and a treeless alpine zone. Between 11,000 and 20,000 plant species have been found in these forests, including a rainbow-colored eucalyptus and two ancient species of the coniferous *Araucaria*. Even the top end of this wide range is likely to be an underestimate. Western New Guinea's flora is largely unstudied, and the east is still steadily adding to the scientific plant catalog. American biologist Andy Mack encountered this grand and comparatively unlabeled flora when he moved to Papua New Guinea in the late 1980s to study how cassowaries disperse tree seeds. He was intrigued by one particularly conspicuous mahogany-like tree whose fruits were a staple for the giant birds. Mack describes the trees as having a glowing appearance. Glow as they might, Mack couldn't publish anything about the bird-tree relationship because the tree had not yet been "discovered" by Western scientists. He recruited a botanist to get out in the highland forest and give the thing a Latin name, which the botanist did, making an unsolicited homage to the big-bird researcher: *Aglaia mackiana*.

New Guinea is a fern champion, with 30 percent of the planet's species. It's also rich in orchids, with 2,850 species, 86 percent of them found nowhere else. There are over 2,000 lichens, of which at least 800 don't yet have Latin names. No plant families are unique to New Guinea, but 60 percent of its species are. Rampant endemism at the species level with none at the higher taxonomic level occurs because the young island inherited all its plant families from Australia and Asia and, in its short but decided isolation, has been busy elaborating new species from those ancestors.

New Guinea is thus named because its people are Black. They looked, to a sixteenth-century explorer named Yñigo Ortiz de Retez, the same as Africans, so he thought it would be a good idea to name the island after a place in Africa. What he missed at first glance is that the dark-skinned people living on this one island spoke over a thousand languages, twice as many as the entire Indo-European family. The forest peoples of the island were nothing like Africans or, for that matter, like each other.

Mid-elevation forest in West Papua.

Five hundred years later, that's still the case. And, even more remarkably, Papua New Guinea's people retain customary ownership of almost all the land, though "ownership" is an awkward reduction of the connection between the humans, the rest of the organisms, the waters, and the mineral substrate atop which they all thrive. Clans' identities are rooted in forests full of sacred sites, ancestor spirits, historical events, and absolutely everything people traditionally needed to exist.

The commercial harassment of New Guinea's plants and animals originally focused on feathers. Birds of paradise are crow relatives costumed in a gorgeous array of plumage, which showed up hundreds of years ago, thousands of miles away, on royal garb in Asia and the Middle East. It didn't take many birds to bedeck the ruling classes, so this trade was not particularly threatening to any of the species. In the late nineteenth and early twentieth centuries, however, feathered headgear became popular among commoners in Europe and North America. The male lesser birds of paradise, greater birds of paradise, and raggiana birds of paradise all court boisterously, waving around foot-long downy sprays colored white, yellow, and salmon, respectively. These, in the collective opinion of fashionable ladies a hundred years ago, looked great on hats. Exporting bird skins became a big business for English, German, and Dutch colonists. North American species, notably snowy egrets, were also being made into headgear. That fired up an American conservation movement and eventually resulted in legislation, a bird-friendly Supreme Court opinion written by Oliver Wendell Holmes in 1920, and a change in fashions.

Since then, generally speaking, people have been taking forest from the birds instead of birds from the forest. Between 1972 and 2002, 15 percent (9.4 million acres) of Papua New Guinea's tropical forests was razed. The damage was done in equal measure by logging and farming, with minor acreages falling to mining, fire, and plantation agriculture. Another 8.8 percent was degraded but not entirely cleared.

Patterns have shifted. A 2015 University of Papua New Guinea report found much less deforestation and slightly less forest degradation in recent years. From 2002 to 2014, the annual forest loss was 75 percent less than the rate reported over the previous thirty years. Degradation fell by around 14 percent. In the relatively petite megaforest of New Guinea, IFLs start out smaller than those on continents and can easily slip below 125,000 acres. Seventeen percent of PNG's IFL area has done so since 2000, well above the global rate of loss.

Most current forest loss in PNG is happening along roads in the lowlands on the northern side of the island and on the country's biggest offshore island, New Britain, home of the volcano-nesting megapodes. Timber, gold, and land for palm plantations all attracted attention from investors in the first decade and a half after 2000. This picture could soon get worse if PNG's plans to expand its road network by over 50 percent come to fruition.

On the Indonesian side of New Guinea, 86.9 percent of "primary" forest (Svetlana Turubanova's definition) remained in 2000. In 2012 it was 86.2 percent. Particularly notable was the 90.8 percent of lowland primary forest still standing in 2012. Forest in flat country is usually the first to go. A separate 2019 analysis published by the Center for International Forestry Research estimated that 2 percent of "old-growth" forest had been lost in Indonesian New Guinea between 2001 and 2018. Oil palm plantations were the leading culprit and consumed increasing acreage starting in the mid-2010s. The remarkably extensive forest coverage in Papua and West Papua is all the more notable because—or perhaps partially explained by the fact that—during this period, Indonesia was deforesting the rest of its islands, overtaking Brazil as the world leader of annual "primary" tropical forest loss.

To date, the most infamous industrial project in Indonesian New Guinea is a complex of mines run by the New Orleans firm Freeport-McMoRan, Inc. The world's largest gold mine and second-largest copper mine, it has operated since 1973, shortly after Indonesia's rule became definitive. The metals dug out of the high Papuan mountains bring in over $4 billion annually, split between Freeport and the Indonesian government. The mine has provided 20,000 jobs—as well as complete destruction of the Aikwa River, armed rebellion, prostitution, high HIV rates, and tens of millions of dollars in suspect payments to the Indonesian military. Freeport's mine is a template for excavation that could one day pock the mountains throughout Indonesia's side of New Guinea.

Looking to the future, plantations are an acute threat on the side ruled by Indonesia, which has a highly developed palm oil sector. Planners in the sinking capital city of Jakarta draw up land-use plans for their faraway islands, based in part on soils, climate, and especially topography. Anywhere that's flat is a candidate for deforestation to make way for large-scale agriculture. In such zones, companies can apply for a concession with the national government. In the Papuan provinces, companies need to reach an agreement with the Indigenous customary landowners; obtain the support of the local government, called a regency (or *kabupaten* in

Indonesian); and satisfy environmental impact review from the province, which is the jurisdiction intermediate between regency and nation.

This may sound like a lot of red tape, but the oil palm magnates manage. As of early 2020, there were 4.5 million acres of oil palm concessions in Indonesian New Guinea. That's when Luhut Pandjaitan, Indonesia's chief minister in charge of investments and one of the crop's biggest cheerleaders, came to West Papua with the surprise announcement that no more palm planting would happen. Farming firms would instead have to cultivate something else, like nutmeg or coffee. Activists were skeptical that the environmental and social story would change; past government pledges to plant non-palm crops have been ignored, and alternative crops can decimate the forest and dispossess traditional landowners just as completely as oil palm can.

In October 2020, a piece of legislation called the Omnibus Law on Job Creation passed, dramatically weakening environmental protections. Mubariq Ahmad is a prominent lifelong environmentalist and outdoor enthusiast from Sumatra. He has a masters from Columbia and a PhD in economics from Michigan State. He's the former head of the World Wildlife Fund in Indonesia and spent several years as lead environmental economist at the World Bank in Jakarta. Now he runs the NGO Conservation Strategy Fund–Indonesia. Ahmad has a buzz cut and a friendly round face. But when we met him in early 2020 in a Jakarta cafeteria, his smile didn't last long.

"Yes, I'm angry!" he said, describing the Omnibus bill, which was proposed the same week we met with him. Ahmad said it was part of a trend to ease forestland into the hands of developers.

"This is my biggest fight with the folks in Jakarta now. The Minister of Agrarian Affairs and Spatial Planning just happens to be my friend from a long time ago, almost forty years ago. . . . He always says that land in the hands of people, especially Indigenous people, doesn't have value. That's why we need to bring in investment. I said to him, 'That is a very colonialist view. You don't want to recognize the rights of the people. How come the same piece of physical thing today, if it is in the hand of the people, doesn't have value, and if the forceful takeover happens, and the next day it is in the hands of the conglomerates it suddenly has value?'" Ahmad has witnessed these takeovers in countless communities elsewhere in Indonesia. Locals lose land, food sources, and the roots of their culture, the land loses its trees, and the planet loses another line of defense against climate change.

The Samdhana Institute is an antidote to all that. The group, based in Indonesia and the Philippines, helps Indigenous peoples keep their livelihoods and spiritual lives entwined with nature. Yunus Yumte leads Samdhana's work in New Guinea and guided our visit. He's a member of the Mare subtribe of the Maybrat tribe in West Papua. He grew up in the West Papuan city of Fakfak, which he thinks is hilarious to tell English speakers. His parents had moved there from their village for school and professional jobs. Yumte attended Indonesia's most elite forestry school, at the Bogor Agricultural University on the island of Java. There he worked his way into projects with world-class experts at the Center for International Forestry Research.

Yumte has a compact build, wide eyes, and a way of inciting his companions into riotous laughter over just about anything. He loves John Denver and swaps "West Papua" for "West Virginia" when "Take me Home, Country Roads" plays. He's also hyperresponsible, carrying two cell phones, a huge black laptop, and various documents organized into plastic folders. We were accompanied by Samdhana's Sandika Ariansyah, a chatty Javan who enjoys easy rapport with villagers and always has his companions' well-being in mind. He carries the first aid kit. Betwel Yewam rounded out the crew. He's a tall, handsome fellow with a shaved head, a deep laugh, and few English words, though he seems to understand quite a bit. Yewam was often barefoot during the journey.

Yewam, a member of the Abun tribe, took us to his home region on the wild, beautiful north coast of West Papua. His village of Kwoor is at the end of the coastal road that rounds the far northwest of the island, passing dozens of deserted tropical beaches. At Kwoor he recruited a childhood friend, Derek Mambrasar, to lead us into the forest via the Kwoor River. As we gathered food, fuel, an outboard motor, and a long section of bamboo

Yunus Yumte of the Samdhana Institute helps Papuan communities sustain customary relationships with land.

full of local coconut home brew, the skies opened and a biblical downpour flooded the streets of Kwoor. It whipped the Kwoor River into a deadly torrent of scissoring logs. We would need to ascend this river to get to our forest camp, so we decided to wait it out in Mambrasar's house for the night. In his main room, he had five big racks of deer antlers, a couple of hunting puppies, some chickens barricaded in a corner, and a tiny python in a bottle. While we waited, he treated John with nettle leaves collected from the forest. Rubbed on bare skin, they produced intense stinging pain, followed by a dreamy topical narcotic experience.

The next morning we embarked at the mouth of the rain-swollen Kwoor River after requesting permission from the customary landowner of the forest where we planned to camp. The forested hills were abrupt and animate, seeming to kneel into the river, pushing it this way and that. A juvenile white-bellied fish eagle soared overhead. A couple of hours inland we pulled out of the current and into a tiny tributary, the Sumi. The little river was obstructed by a fallen tree, so we unloaded and trekked through the welcome shade of the woods. Fallen leaves, colored in chestnut and coffee tones, were still wet from the overnight downpour. The forest smelled like opening a cabinet with bitter chocolate inside. After 15 minutes, we rejoined the creek on an ample beach across from a stand of feathery rattan. A freshet of drinkable water issued from the forest slope.

We set up our camp and gathered banana leaves for seats while Yumte whipped up a meal of cabbage, carrots, ramen noodles, and canned fish. In the late afternoon, Mambrasar mustered the group for a walk to see forest that had been exploited for the prized merbau timber, known in English as ironwood (*Intsia bijuga*), as well as adjacent areas where a landowner had turned down loggers' offers. We set off walking in the shallows of the clear Sumi River.

We crisscrossed the creek, climbed banks and traversed bits of forest, and then splashed back into the water until we came to the logged area. The damage was subtle. The scattered stumps were weathered and saturated to a deep dark brown. Clearings opened by treefalls had thickened with shrubs and small trees and required a bit of machete work to negotiate. A logging road, overcome by regrowing forest, nearly escaped our attention as we crossed it. After a while, Mambrasar led us out of the logged zone and back to the river.

John noticed quite suddenly that his right knee and thigh had huge painful welts blooming. Betwel Yewam looked at the leg, concerned. "Itchy leaf," he pronounced. He took hold of the leg and raised his machete for one terrifying beat and then whacked the afflicted spot with the flat of his blade. John was shocked momentarily and then delighted by this swift jungle remedy—until Yewam advised him that it would stop hurting in a couple of days (which was accurate). Yewam pointed out a specimen of the offending plant, a nettle that looked similar to a great many other plants in the forest, innocent and leafy.

The intact, unlogged ironwood forest we saw next had a pleasant, airy understory and no stumps. Many trees exceeded 3 feet in diameter, with both fluted and round formats. These old organisms were still in the forest because of their relationship with a clan of humans that, so far, had decided that trees are better than money. It made one wonder if the trees know—or if they live blissfully unaware of the possibility of becoming garden benches, doors, and fancy dining tables.

As we climbed a slope, it started to rain. We emerged on an old timber haul road that had not recovered like the smaller logging tracks we had seen earlier. A deer grunted from the bush. Mambrasar explained that the Timor deer, introduced to New Guinea by the Dutch from other Indonesian islands, browse the road, helping to keep it treeless. "*Kampung rusa*," he said in Indonesian with a grin: this is the deer village. Our group moved across grass, gravel, and spiny thickets at a near trot in the gathering twilight.

By the time we turned off of the road and back into the trees, it was fully dark. Our headlamps spotlighted ovals of forest. The ground was soft, with slick exposed roots and occasional creeks cascading in the hillside's deeper folds. The air smelled like earth. We descended, gripping the ground with toes and grasping one skinny trunk after another, careful not to get a handful of millipede. They were 6 inches long, fat and glossy, scaling the slender trees in a nocturnal commute of some sort. Mambrasar kept up such a brisk pace that there was no time to ask about them. The dark and the challenge of staying on the heels of a forest native who's darting through the woodland home of uncounted generations of his ancestors induced a sort of mindless focus. The night felt like the start of something.

We didn't see much that appeared dangerous. There were no tarantulas or columns of aggressive biting ants. Anyone who has inadvertently grabbed a tree house of South American fire ants or a spine-encrusted palm in Central America looks before handling a tree. Deadly vipers, common in many jungles, are, Mambrasar said, not a threat here. A biologist friend later questioned this, saying that snakes called death adders are "quite common" in this part of West Papua. But it's possible both things could be true. The snakes—and other biting, stinging things—could be quite common and also not much of a threat if one is walking behind someone like Mambrasar on his home turf.

We stopped by a tiny stream to refill water bottles, smoke, and listen. A rufous owl cried in the distance. Another animal sound made all of us visitors look around at one another. It was a beep like a gadget running out of batteries might make. Water frogs barked like German shepherds, and smaller, unidentified frogs yapped like chihuahuas. As we resumed our walk, a new apparition, looking straight at us from about 25 feet away, put the New Guinea megaforest emphatically in its Australasian biogeographic context.

A kangaroo.

On Mambrasar's command, the rest of us hit the deck and extinguished our lights. Gaping at the marsupial irises shining in the guide's beam, John internalized for the first time that New Guinea has kangaroos. He knew about the cute orange tree kangaroos. But this animal, which Mambrasar identified as a *kanguru* in Indonesian and in English officially is called a forest wallaby, is a big, bona fide, jumping around sort of kangaroo.

By any name, it's meat.

Mambrasar dropped his string bag and, toting a spear he had fashioned from a sapling, advanced. He played his flashlight in the canopy, spotting randomly to bamboozle the animal. The idea was to get the kangaroo to focus on the forest ceiling rather than the man. In many, perhaps most,

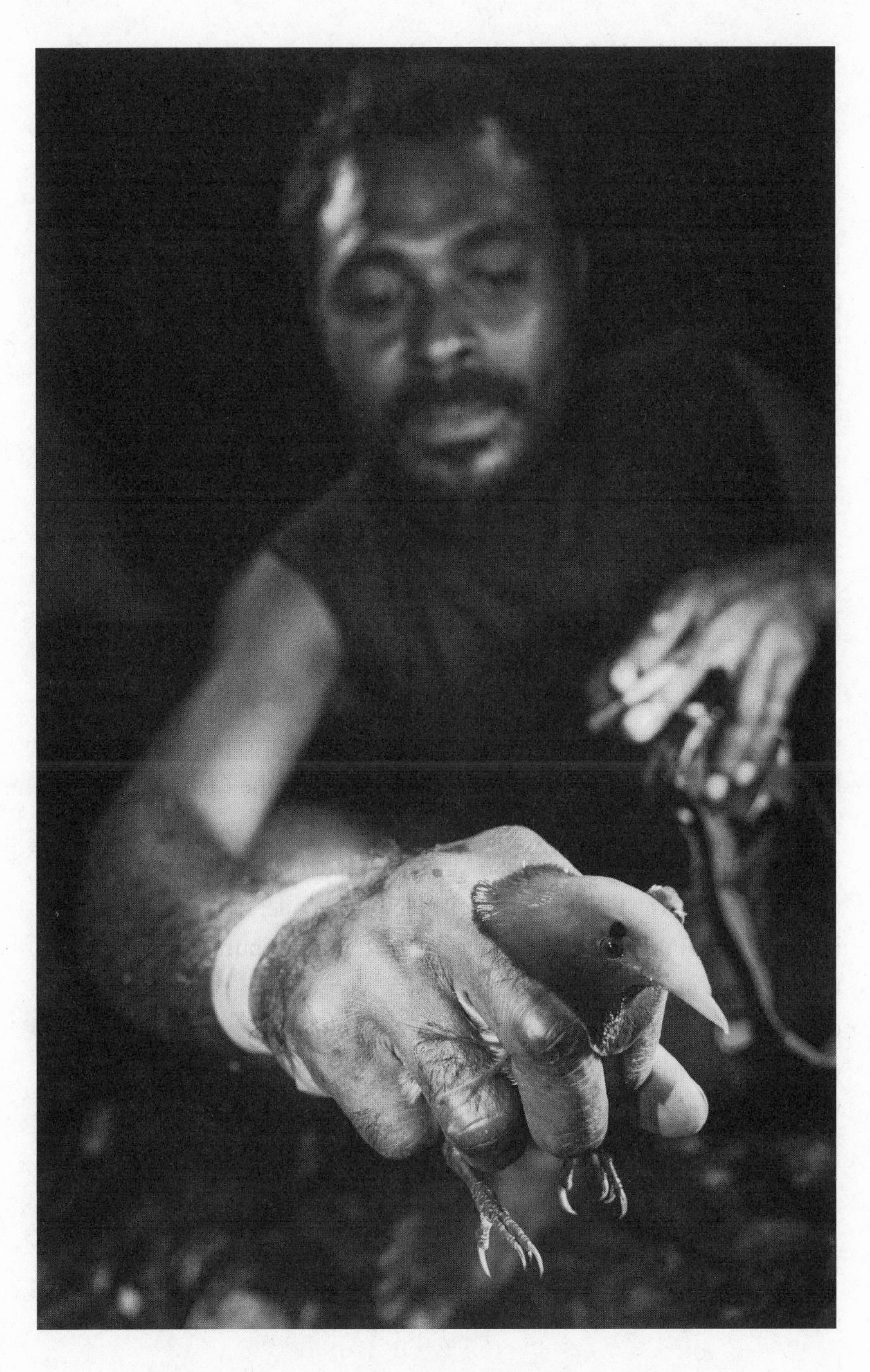

Derek Mambrasar and a king bird of paradise in West Papua.

forests worldwide, Indigenous people have been quick adopters of firearms. Many still know how to hunt with blowguns, spears, arrows, clubs, and nets, but, given the opportunity, are not immune to the charms of a hunting tool that permits a person to kill things from far away with terrific accuracy. Indonesia, however, does not allow its citizens to possess firearms. In Ayapokiar, where the Momo clan lives, most boys and some girls run around with bows and clutches of small arrows. To fish, kids employ arrows fitted with barbed trident points. Adults carry spears or bows in the forest just as people have done for tens of thousands of years. The Indonesian Army and police, however, are not traditionalists. Some use their spare time and ammunition to run side businesses in Papua and West Papua slaughtering deer, pigs, and kangaroos.

Mambrasar briefly confounded the kangaroo; it hopped this way and that on a stage of flat ground in what looked like a dance with the hunter. Presently, however, it hopped over a rise and into the depths of the forest.

Shortly before the end of our 4-hour circuit, Mambrasar spotted a king bird of paradise. The bird was asleep. The man had no qualms about interrupting. Mambrasar trotted over and—with lit cigarette in one hand—leaped like he was grabbing a rebound and snagged the *Cicinnurus regius* specimen off its perch and out of bird dreamland. It was a scarlet-and-white confection with iridescent turquoise wing tips, electric blue feet, and hazel eyes. The tail feathers looked jeweler-made: twin wires 6 inches long that terminated in coils, with emerald tops and spun gold on the bottom. After a few minutes, Mambrasar let it go. This matter-of-fact bit of catch-and-release bird-watching brought home just how fully and physically involved Mambrasar, and forest people like him, are with the world we call nature.

5

Forests of Thought

Megaforests hold staggering human diversity. Over a quarter of Earth's languages are spoken in the world's largest woodlands. The mere tally of languages, however, is less arresting than their particulars. Thousands of lexicons are deployed according to grammars that seem to test every possibility of human perception and cognition. They throw open the conceptual boxes within which each of us thinks and, in doing so, reveal the full spectacle of human inventiveness.

The ethnobotanist Wade Davis calls each language "an old-growth of the mind." In his biography of his mentor Richard Evans Schultes, Davis describes Amazonian peoples Schultes met who use the same word for green and blue—a single color hinting at a connection between the two overhead vaults, one of trees and the other of sky. Meanwhile, these people had eighteen different terms for ayahuasca varieties indistinguishable to Schultes, who lived in the rainforest for over a decade and was one of the most accomplished tropical botanists of all time. Megaforests have spawned thousands of cultures, each with its own unique way of perceiving reality, processing information, and making it into verbal expression. Each, in other words, with its own way of being in the world. Now, intact forests shelter this diversity in a world where at least 43 percent of languages are endangered.

While language doesn't equal culture, linguistic diversity provides a measure of the variety of cultures that persist in different corners of the planet. The most complete database of world languages is Ethnologue, which contained 7,139 entries as of early 2021. It sits behind a paywall at the American missionary organization, the Summer Institute of Linguistics, now rebranded SIL International. SIL missionaries started learning Indigenous languages in the 1930s in order to live with peoples who, according to SIL, "needed language development," which was code for translating the Bible into (well-developed) local languages. SIL's evangelizing zeal has powered a global map that features a tiny black dot for the approximate center of each territory where a distinct language is spoken. The dots are scattered like poppyseeds across a world map. The island of New Guinea looks as if it was dropped wet into the poppyseed bin; it's almost entirely black with dots. With a blown-up Ethnologue map that is included with the $2,400-per-year subscription, one could count over 1,000 in total, 851 of which belong to Papua New Guinea, the island's eastern half. New Guinea language expert Bill Palmer, at the University of Newcastle in Australia, counts 1,300, including some spoken on smaller islands immediately to the east and west of the main landmass. To put New Guinea's extreme linguistic diversity into perspective, geographer Jared Diamond attempted a reckoning of how many languages existed in several regions shortly before 1492. For every 1 million square kilometers (386,000 square miles), he estimated that Europe had 6, Africa had 49, California had 156, and New Guinea had 1,250.

New Guinea is the global epicenter of language diversity, with an extravagant array of families—at least forty—and copious languages within them. Families are groupings of languages evolved from the same source and diverge mainly in vocabulary—Spanish and Portuguese, for example. Different families are dissimilar in their grammars, like Spanish and Chinese. New Guinea also has numerous languages known as isolates. Like the European isolate Basque, these are one-language families whose lineages haven't been linked with that of any other tongue.

One New Guinea isolate is Maybrat. It's the native language of Yunus Yumte and also of Fince Momo, whom we visited in the Tambrauw Mountains of West Papua. Sitting around on mats made of pounded and sewn leaves, we tried to learn a little Maybrat while Momo fried hunks of fish freshly caught from the nearby Iri River. The pan of bubbling, hissing oil sat atop three sticks angled into the ground. The tripod was expertly positioned so that it burned just a bit more slowly than the food cooked. Momo's niece Anastasia played with a big spoon on the forest floor. We asked them how to say "forest" in Maybrat.

It depends. To describe the forest we were sitting in, you start with your tongue pressed against your upper teeth and blow a "toe" sound out, then finish by exhaling a long "fff." It's written "*toof*," but sounds more like "toeff" and feels like clearing water out of a snorkel. *Toof*, Yumte explained, "is the forest that people are touching very often and someday they can convert into traditional garden or crops. They explore it for hunting and collecting." To a visitor's untrained eye, there was no obvious sign of people, and the path that led here was insubstantial, dwindling in places to a faint thread of subtly trampled leaves. As unbroken and intact as this forest appeared, it was a zone actively used by the people of Ayapokiar. In fact, we sat at a place in the forest with a specific name, *srokha*, which means "hangout spot" or "meeting place," because it is an important woodland crossroads.

If we were in a less "touched" forest, most Maybrat speakers would call it *arin* and a few clans would say *atrouw*. Yumte ventured a translation to the inexact English term, "primary forest." Then there's a rarely used name for secret, sacred forest places, *moss*. The term can also mean "nature," in the sense of the whole universe. Yumte added that in this forest-covered land, forest is also called "land," which is *tabam* in Maybrat. There's no specific term for the English-language concept of "nature" as a separate nonmanufactured realm that includes settings as diverse as forests, grasslands, and coral reefs. That's a new idea to the tribes who speak Maybrat.

If West Papua's woods become splintered and their cultures uprooted, the Maybrat ideas of forest—*toof*, *arin*, *atrouw*, and *moss*—and the *nden* of the neighboring Abun people, who speak another linguistic isolate, risk fusion into the Indonesian "*hutan*," a "forest" with no local roots or

Fince Momo in the *toof* in West Papua.

resonance. Forest peoples will be left with a splintered forest and compelled to talk about it in a trade language of seafaring Malayans, a tongue that is Indonesia's lingua franca thanks in part to its lack of intricacy. Each lost language narrows our view—not only of forests, but of the world. There is an ethical imperative to avoid the identity-wrecking loss of megaforests and also pragmatic sense in refraining from burning libraries of forest knowledge just when our planet's stability needs them most.

Some speculate that New Guinea's languages are diverse because people have been there for a long time: at least 47,000 years and perhaps as long as 65,000. The original human settlers have had ample opportunity to disperse throughout the island, occupy all sorts of isolated nooks, and begin talking differently. Subsequent waves of newcomers, such as Austronesians originally from the island of Formosa (now Taiwan), periodically showed up, spicing up the language mix. But just as time can diversify, it can also homogenize, particularly when a dominant, imperial tribe emerges.

Physical obstacles offer another explanation. Central New Guinea has an accordion topography, bounded by thick, river-laced, and sometimes marshy jungles along the island's coastal zones. Rivers and wetlands also impede cross-country travel in the language-rich Amazon and Congo lowlands. The single-most diverse language region anywhere—like a kaleidoscope within a mosaic—is the Sepik River basin, a swampy area of northwestern Papua New Guinea. If it's hard to get from one place to another, the thinking goes, people don't bother. They don't interact and thus end up speaking different languages. This is a partial explanation, though not without counterexamples, such as the Incas and Maya, who mastered formidable mountains and jungles to assert cultural hegemony.

Forests help complete the picture. Daniel Nettle, a British anthropologist at Newcastle University in the United Kingdom, crunched rainfall and language numbers in the tropics and found that wherever it rained a lot year-round there tended to be more languages. He concluded that people who live in places with plenty of consistent rain and stable temperatures throughout the year can subsist in isolation; they don't need to forage in different territories or trade routinely with neighbors to get through dry or cold seasons. Nettle's findings have been validated by more recent statistical research, which teased apart whether isolating terrain or precipitation is more closely associated with language diversity. A monotonously wet climate emerged as the strongest predictor. Because photosynthesis requires water, watery places are also prolific in the conversion of carbon dioxide gas into biomass. These ecosystems where languages and cultures multiply are also those where the vegetation thickens and grows tall—forests.

The most dramatic examples of unique human cultures harbored by big forests are the so-called uncontacted peoples, self-contained societies that live entirely from their forest environments. Nearly all of them are in remote parts of the Amazon. Formerly the stuff of rumor and legend, 185 groups have now been preliminarily detected in South America, of which 119 have been confirmed with aerial surveys or very careful ground expeditions. They ring the basin, far from roads, along streams too narrow or rocky to navigate. Most live in the forests identified in a 2020 *Science Advances* paper as the world's two largest unbroken expanses of tropical woods. People are able choose this life because there are still intact forests.

A large measure of credit also goes to Sydney Possuelo. For over twenty years, until 1987, Possuelo's job for the Brazilian government was to contact uncontacted people. Brazil's military rulers were pressing into the Amazon and considered contact inevitable. Possuelo was a *sertanista*, a term derived from Brazil's arid northeastern *sertão* region, which, before the Amazon was explored, was the wildest place "civilized" Brazilians went. *Sertanistas* like Possuelo would plunge into the Amazon on long expeditions to locate uncontacted people and coax them into interaction by offering prized items like metal tools and cookpots. Once thus persuaded, the newly contacted group would become wards of the federal government and the Catholic Church. Many, sometimes most, would die of disease. Those who survived would live as Brazilian Christians in misery, marginalized and often alcoholic. After one particularly heartbreaking episode, with the Arara people near the Trans-Amazon Highway, Possuelo fully realized that his society had made *sertanistas* their angels of death.

In 1987, as head of the government department dealing with uncontacted tribes, Possuelo changed Brazil's policy from seeking contact to avoiding it. This radical change coincided with Brazil's rewriting of its constitution, approved in 1988, which recognized Indigenous peoples' right to exclusive use of their ancestral territories. Possuelo and colleagues demarcated hundreds of territories. Many were inhabited by both contacted and uncontacted peoples, including the two largest Indigenous territories in the country—the 22-million-acre Yanomami land and 20 million acres of the Javari Valley.

A handful of recently contacted tribes have successfully opted to maintain pre-contact ways of life thanks to the insulation their mega-forest homelands offer. The Zo'é, for example, were contacted in 1987 by American evangelicals from the New Tribes Mission in cooperation with the Brazilian government. At the time, the country's military leaders were planning a road that would traverse Zo'é territory in the northern Amazon. The Zo'é tried out the missionaries' gifts of clothes and firearms and listened to the gospel. The clothes made them itch. Guns required ammunition that they could not fabricate themselves and also made it too easy for them to kill a lot of animals at once. The weapons would require new social controls to avoid wiping out all the game within easy walking distance of their villages. They rejected these innovations, and the gospel, too. With Possuelo's help, they banished the missionaries in 1991.

Thirty years later, the Zo'é hunt, heal, farm, and govern themselves according to their traditions in a legally designated 1.65-million-acre territory embedded in the world's largest contiguous tropical forest, which is over 170 million acres (crucially, the road was never built). The Zo'é wear almost nothing other than a wooden ornament called a *poturu* in their pierced chins. The modern contrivances most popular with the Zo'é are watches and written language. Secular linguists have helped to create written Zo'é (a Tupí-Guaraní language) in which the tribe has now penned a long-term plan for their land.

Zo'é is one of the Amazon's 350 languages, which belong to six major families with twenty to eighty languages each, plus at least a dozen smaller groups that have up to twelve languages. Language family members are geographically scattered rather than neatly clumped, leading one expert to liken the language map of the Amazon to a Jackson Pollack painting. The basin also has scores of isolates. And that's just what's left after centuries of interaction with Europeans. According to linguist Alexandra Aikhenvald, a professor at the James Cook University in Australia and author of

The Languages of the Amazon, pre-contact languages likely numbered 600 to 1,200.

The diversity of verbal expression in the Amazon has bewildered scholars seeking a universal grammatical principle that governs how humans turn thinking into speech. The idea, a linguistic "Holy Grail" championed by Noam Chomsky, is that cerebral hardwiring constrains the way our species can make language. The Amazon megaforest is where such universality theses go to perish. R. M. W. Dixon, Aikhenvald's colleague at James Cook University and one of the world's eminent language experts, wrote in 1999 that he had "devoted several decades to searching for the substantive linguistic universals. In case after case, just as I thought I had achieved some significant typological statement, a counter-example popped up; and this was invariably from a language of Amazonia."

Amazonian languages overturn some fundamental assumptions that speakers of European languages may hold. In many Amazonian languages, there's no phrase for "thank you," for example, or words for numbers above three. And yet, gratitude exists, and quantities have been managed for millennia before infinite base-ten counting arrived from overseas. Many in the region have no word for "have" (the second-most frequently used English verb). After contact with speakers of Spanish and Portuguese, Amazonians often adapted their word for "hold" to cover this new idea of a possessive verb. The possibilities of possession vary by tribe. Pronouns such "my," "our," and "your" cannot be applied to the sun, moon, or jungle in Tupi-Guarani languages; to rainbows, deer, or kinfolk in Macushi, along the Brazil-Guyana border; nor to manioc or water among the Hixkaryana, who live on the Mapuera River in Pará. Land is linguistically unpossessable for the Baniwa, a tribe on the Rio Negro.

Many Amazonian cultures prize epistemological precision, mistrusting people who fail to provide the details of how they know something. So, most languages of the region fastidiously footnote verbs by adding "evidential" prefixes and suffixes that specify whether the speaker saw something happen, apprehended it nonvisually, heard about it, inferred it indirectly, or found out some other way. A Matsés speaker, for example, may be certain that a recently deceased person with no obvious signs of trauma has been killed by a shaman but will say *"nënëchokid-n akak"* only if she has seen the corpse. If she didn't see it but is still sure, she'll use what's called the conjectural form of "to kill" and say *"nënëchokid-n akash."* The Tariana, Tuyuka, Desano, Tucano, and deep-forest Hup and Yuhup peoples have grammatical tools to indicate five categories of evidence. A related tendency of Amazonian speech is to retell events in full dialogue rather than paraphrasing the characters involved.

The Pirahã are among the most isolated and linguistically exotic of tribes. They live south of the Amazon River and have an elaborate gender system that includes pronouns specific to terrestrial and aquatic animate nonhumans. Where in English "it," "he," or "she" might be applied to monkeys and fish, these animals get their own pronouns in Pirahã that sound something like "ik" for the former and "si" for the latter. Their language is sung more than spoken and has a globally unique sound, a kind of "g" that requires a maneuver linguists call a double slap. In this case it involves tapping the tongue tip against the alveolar ridge—the bone that holds the upper teeth—then sticking it out and smiting the lower lip.

In contrast to the other two tropical megaforests, the Congo basin underwent major pre-European colonization events. Starting around 4,000 years ago, Niger-Congo-speaking millet farmers from West Africa migrated southeast, settling in the Congo basin. They were followed by a wave of iron-working farmers. Other migrants came from Northeast Africa, bringing tongues of the Nilo-Saharan family. The farmer-migrants' languages supplanted vernaculars of the forest peoples they encountered in the basin. Niger-Congo became the world's largest linguistic family with 1,524 entries, as the migrants' languages differentiated richly throughout the Congo and into southern and eastern Africa. The DRC alone has over 200 languages, while Gabon has 40 and the Republic of the Congo is home to 60.

Despite the presence of dominant "newcomers," the Congo's original forest cultures persist. Three of today's twelve recognized Pygmy peoples still speak their own languages, which have grammars and some vocabulary from the Niger-Congo and Nilo-Saharan languages, plus rich lexicons of forest-related terms that are distinct and possibly derived from their pre-contact forms of speech. Through 150 generations of Black colonizers, as well as White powers that appeared 150 years ago, Pygmies have retained a unique status as the people who know the forest best. Few non-Pygmies will venture there without them. In the Goualougo Triangle, John learned why.

After a morning watching the co-feeding apes, he went with Goualougo's camp manager, Sean Brogan, and two Pygmy trackers, Emilie Bakombo and Ndambio Justain, to see a wetland clearing in another part of the forest. On the trail, safety protocols require that one tracker always be at the front of the line and one at the rear—always paying attention. A half hour into the walk John was walking several paces behind Justain when suddenly the young man froze. Without looking back, Justain stretched his arm out behind him and urgently opened and closed his hand. Did it mean "Stop right where you are," "Turn around and run like hell," or "Come here and check out what I'm seeing"?

John chose option 3, which was wrong, because what the young Pygmy saw was a forest elephant. These elephants are smart, very big, run faster than humans, and have a long score to settle with us; their population has been reduced by 90 percent due to the ivory trade. There is no scientific evidence that forest elephants bear grudges, but we do know that they mourn and have long memories. In a nearby research camp on the Mbeli River, there is a shrine that attests to the danger the animals can present. The memorial is for a young Danish scientist killed by an elephant in 2018 while she was walking in the forest without Pygmies.

A mere 25 feet away, sunlight had found a gap in the canopy and illuminated the animal John would have walked right into had he not been preceded by Ndambio Justain. She was radiant. She showed a long, subtly curved left tusk, an unflapping ear, and a corrugated trunk dangling perfectly at ease. The right half of her face was concealed by a spray of leaves. She looked as though she were lost in thought.

Justain turned and we ran like hell, following Emilie Bakombo. The sprint was primal, running not for exercise or fun but to escape real physical peril. Eventually we stopped, and the two men conferred. They discussed the animal's probable next move, pointing this way and that into a forest that, however inscrutably uniform it might appear to a visitor, for the two trackers was varied and readable. We set off again into the macramé of trails, successfully avoiding the elephant and arrived 15 minutes later at the wetland.

Congo forest people seem to perceive in full color and redolence, and with senses beyond the traditional five, a forest that to the rest of us is comparatively odorless, mute, and monochrome. Researchers all have stories about uncanny feats of Pygmy pathfinding and sensitivity to the nonhuman world. Richard Malongo, a Congolese biologist who is the head of the WCS program in the Republic of the Congo, tells of a research project for which he had been making daily trips to a clearing frequented by bongos, a species of orange-and-white-striped antelopes. One morning in camp, the Pygmy tracker working with him announced that on that day they would see more bongos than ever before. The previous record was nine. That day they saw twelve. "It's like he had a dream," Malongo marvels. On another occasion a Pygmy tracker suddenly stopped Malongo's group and announced that they would have to wait because there was an elephant up ahead. No one else had heard or seen a thing. That was because, the tracker explained, the elephant was asleep; he could tell by the smell.

Gaston Abea has the distinction of being the only Pygmy leading his own research with Dave Morgan's ape program. He follows a group of gorillas led by a silverback named Buka every day and studies a behavior that's only recently been recognized by biologists. "People saw the gorillas scratching the ground and putting something in their mouths and assumed they were eating ants," said Abea, demonstrating by scratching at the earth as we sat on stumps at dusk in the research camp. A subtle fungal aroma teased by Abea out of the jungle's bouquet of fruits, buffalo dung, ape musk, and honey tipped him off several years ago that the gorillas were not eating ants. "*Champignons*!" he exclaimed in French. "They were eating truffles." Mushroom eating by gorillas isn't entirely new to science, but only sketchy details were known before Abea started documenting the techniques and patterns of Buka's group.

Gaston Abea detected a rare diet choice—truffles—in a group of western lowland gorillas.

Later in the evening at camp, a dozen of us chatted around a fire that sent sparks up into a puzzle piece of starry sky. Over the fugue of a billion insects, one of the trackers, Eteko Lambert, plucked a tune on a *mondume*, a sort of triangular, four-stringed harp played atop a large aluminum cookpot for resonance. He sang in a high, sweet voice, fingers flying on the strings. When he finished the 20-minute song, he explained that it was about the contentment that results from being in the village with family and friends, sitting around outside, with children running in and out of the forest.

Across the northern megaforests, Indigenous peoples, while not entirely isolated, have been partially spared from the homogenizing tsunami of mainstream culture that has swept through the temperate regions thickly settled by migrants. Canada has at least sixty Indigenous languages in thirteen families. Alaska adds another dozen languages. In Russia, the government recognizes forty "small-numbered" Indigenous groups, defined as peoples with less than 50,000 members. Most of these unique cultural groups live in Siberia and the country's Far East. There are several larger Taiga Native peoples, over the 50,000 population threshold, notably the Sakha and the Buryats. Many native boreal languages and cultures have survived despite brutal repression and assimilationist tactics on both sides of the Pacific.

The Ross River Kaska Dena community is an example. Elders John Acklack and Clifford McLeod took us out into the snowy woods one day to explain how their families formerly made an annual "round," a sort of food tour of the territory, pursuing fish, birds, mammals, and plants—many of which still feed and heal people. Over the course of a couple of days, John had been pursuing a more generalized question about the significance of "the forest" in Kaska culture. The elders gently deflected his queries. And as we gazed around at the landscape mosaic—of spruce, aspen, fir, willow, lakes, swamps, recovering burns, and entirely treeless mountaintops—the view made the idea of a "forest" you could draw lines around and discuss seem elusive.

We stopped and built a fire, roasted some sausages, and shared a thermos of tea. John had one more go at a "forest" question with help from Josh Barichello. Barichello, in his mid-30s, wore cop sunglasses and wild blond hair barely contained by a homemade cap with his name knitted on it. He has spent much of his life around the Kaska. As a child he accompanied his father, Norman, who has studied gyrfalcons in and around Kaska territory for decades and has been a longtime adviser to the First Nation. Now they are both helping out on the Ross River Kaska's proposed Indigenous protected area. The younger Barichello has lived in Ross River since his early 20s, working in the tribal land department and spending a lot of time with the elders compiling their traditional knowledge. Josh is one of the few young people in the community who speaks Kaska, which, like Navajo, is a member of the Athabaskan linguistic family.

John's doggedness asking about "the forest" in local culture now provoked a joke in Kaska from the elders. Acklack and McLeod had a good laugh and Josh Barichello grinned. Later, Barichello translated as he guided our four wheeler back through the trees. The joke referred to a story involving Wolverine—a character with more energy and persistence than good judgment—and two lovely maidens who end up in the canopy of a tall tree with an aroused and stymied wolverine at its base.

Back in the village in the evening, Acklack and McLeod served up a medicinal tea made from balsam fir bark collected on the outing. They insisted everyone have some. As we sipped, Josh Barichello ran a Kaska translation for "forest" by the elders. He had been ruminating on the question of how best to put the English idea in Kaska since John had asked the day before. "Thickly spruced" is the first term Barichello had come up with. Now the three discussed the matter and agreed on an additional way to say it: "*dechin tah*."

This literally means "among the sticks." The English noun "forest" is a Kaska prepositional phrase. Forest isn't an object; it's a situation, a phenomenon brought about through a relationship involving a person and a place. After a while Acklack and McCleod headed for home, carrying hot

Balsam (subalpine) firs in Kaska Dena country.

jars of the fir tea wrapped in brown paper, like something precious. Like an answer.

It's easy to lose sight of how short a time people have been in a position to ignore wild plants like balsam fir and not perish within a week or two. For most of human history, knowing the chemical and physical properties of what grows around us was necessary for shelter, health, food, fortune, vision, romance, personal ornamentation, altered consciousness, and logistics. Intact megaforests preserve old forest science that has accumulated over extremely long experimental periods. Jess Housty is a young

Kaska elders Clifford McLeod (left) and John Acklack relax in a four-wheeler.

practitioner of this old science. She rambled the big woods, beaches, and boggy muskeg of coastal British Columbia as a child. She's now a leader of her Heiltsuk people. First elected to the tribal council at the age of 26, she was about to finish up her second four-year term when we spoke in March 2020. She is also executive director of the Qqs ("kucks") Projects Society, which incubates programs that support Heiltsuk youth, culture, and environmental protection. When still a child, Housty's adopted grandfather, a man named Ed Martin, told her she would learn about plant medicine. He wasn't musing or suggesting. He was announcing Housty's obligation.

"Growing up, my grandparents would point at the shoreline and the ocean and the forest and say, 'there's your medicine cabinet.' That's where we'd go for the things we need to heal each other. That was a big part of growing up. We didn't go to the doctor when we were sick. We almost always were able to treat the things that needed to be treated through the things we harvested. I have a box under the spare chair in my office that's full of jars of different medicines in case people come by and need something." She rattled off a list: Labrador tea and juniper berry, different lichens, ferns, spruce tips, cedar, devil's club, and alder bark. Her 4-year old son Noen collects plants with her and, when afflicted with a cold, will climb up on a chair to reach the spruce tips down from the freezer.

Housty's people combine old science with new. Her brother William exemplifies the mix, with a formal degree in Western natural resource management, plus deep training in traditional songs and histories tied to the territory. He has led research on grizzly bears, involving DNA sampling and complex statistical analysis to understand the health of the bear population and its interactions with salmon and humans. The bears have been seen by the Heiltsuk as forest "gardeners" who distribute salmon into the woods, long before outside scientists traced Pacific coastal forests' nitrogen back to the ocean. The southernmost spot where *Ursus arctos* bears congregate in large numbers to feast on salmon is on the Koeye River in Heiltsuk territory.

William Housty's 2014 paper on the Koeye grizzlies in *Ecology and Society,* though thick with biology jargon, is dramatically different from standard fare in such journals in its focus on something called *Gvi'ilas.* That's Heiltsuk traditional law, a set of precepts handed down orally over the generations. Housty and his non-Heiltsuk colleagues enumerate six Gvi'ilas principles that guided all aspects of the study—from the noninvasive method of collecting grizzly hair samples to the geographic application of the results to the wider territory, all of which is "home" in the same sense that their dwellings are. Heiltsuk leaders note that the Gvi'ilas code in English conveys a pale shadow of its full "multidimensional" meanings, which are tethered to every forest and fjord in the region. But it's a good start toward merging knowledge and ethics that have been earned over mind-boggling time scales with the reductionist rigor of the scientific method.

While Jess and William Housty took on obligations to steward the traditional knowledge of their own culture, Tamasaimpa* was obliged by his tribe to learn the ways of an encroaching culture of foreigners: Brazilians. He is a Marubo Indian who grew up in a remote village called Kumãya, in the Javari Valley. When he was 17, his people decided they needed young members who could handle relations with the outside world, so they sent him to the nearest town, about a five-day walk through the jungle, to learn Portuguese. His teachers were Catholic missionaries and *Conan the Barbarian* comics. After several years of education and a year in the army, Tamasaimpa returned to the Javari in his early 20s and has been fulfilling his obligation ever since. He has been defending the territory, with a canny sense of both adversaries and allies from various strata of the surrounding society.

Much of that time he worked for the federal National Indian Foundation (FUNAI) protecting uncontacted peoples, including a people called the Korubo. Several Korubo groups have emerged from the forest between 1996 and 2019. In 2015, twenty-one Korubos were captured by a neighboring tribe that had experienced their own first contact in the mid-1970s. The two tribes have been rivals for generations and live in uneasy proximity in adjacent watersheds. After tense negotiations, the captors handed over the Korubo to FUNAI. It was Tamasaimpa's responsibility to make sure the newly contacted people survived their initial encounters with the bizarre and disease-ridden modern world.

Tamasaimpa, who has spent most of his life in the forest, was astounded by how intimate Korubos were with it. He marveled at their spatial

* In Marubo culture, adults typically assume single names derived from the name of a child (that is, So-and-So's Father). Most also have names used in wider society, usually a Portuguese first name and a surname that is the name of their ethnic group. Tamasaimpa, who is generally known as Beto Marubo outside his community, specified his Marubo adult name for use in this text.

A *maloca* (longhouse) in Tamasaimpa's village. *(©Sebastião Salgado)*

awareness. "One time I was hunting monkeys with them. I had an American Ruger rifle with a scope to hunt animals way up in the trees. So, there I was with my scope and rifle with fifteen rounds and all the Korubos with their blowguns. They would let loose with a bunch of darts way up in the trees and then wait, listening to where the monkeys fell." Tamasaimpa imitated a Korubo, with a zen-like look on his face. The plant-based curare venom takes a few minutes to incapacitate the primates. He held up a finger and looked toward the ceiling: "Dohff! A monkey falls and the guy says, 'That's not mine, it's my cousin's. Mine? Waaait' . . . Dohff! 'There's mine!' And he knows just where it is! You get it? They have this unbelievable sense of space. Imagine, twenty men out shooting darts up into the canopy all over the place, how on earth do they know whose monkey is whose?" It matters, because a Korubo who eats his own kill is condemned to a long string of bad luck.

This spatial genius applies at large scales, too. The Korubo still have navigational faculties that most of us have outsourced to paper, circuitry, and plastic. "The Korubo has a sense of topography, of geography, that's unreal. It's like he has sonar. He knows exactly how many hills, how many big trees on the left and on the right, how many creeks there are in any direction for huge distances," Tamasaimpa explained, shaking his head in admiration.

The tribe is known for killing with clubs rather than the bows and arrows traditionally favored by all their neighbors. The Korubo use blowguns for smaller game, such as monkeys and birds, but for wild pigs, tapirs, and combat with other people, they wield their signature *bordunas*. This has won them the Portuguese nickname *caceteiros,* which literally means "deliverers of blows," but has also been translated as "head-smashers." These weapons are around 5 feet of dense wood made from *pupunha* or *paxiúba* palms, with a spatulate top like the end of a canoe paddle and an abrupt point on the bottom. The fleet Korubo chase the pigs and can dispatch them with a thrown *borduna* that twirls end over end like a hurled dagger.

Caring for the Korubo during their contact ordeal of 2015, Tamasaimpa became especially close to one elder named Pëxken (PUSH-ken). Their languages are both in the Pano linguistic group; Tamasaimpa, a natural polyglot, quickly learned to communicate with his new Korubo friend. The two hunted together and shared knowledge about their colliding worlds. "His interaction with the environment is incredible," Tamasaimpa said. In an attempt to get this across, he held his hands out in front of him, fingers spread, palms toward each other, as if he was about to clap. One hand was Pëxken and the other was the environment. Tamasaimpa cast about for a way to explain. Slowly, he brought his fingertips together. "He and the environment are the same thing. The tree *was* his hand."

The Korubo greet visitors by forming a circle, grasping hands, and stomping in a slow counterclockwise circle while chanting *hey-hey-hey!* One man sings a high solo over the drone. Protocol is to lock eyes with the person next to you for the duration of the dance, which lasts several minutes. The practice gives them a good long look at a newcomer. Tamasaimpa says that the sentiment behind the chant is, "I want to see your soul." They don't always like what they see. The Korubo permanently banished one of Tamasaimpa's FUNAI colleagues for something they saw in his eyes.

Sitting on the floor in a Korubo *maloca* (longhouse), the first thing you may notice is all the holes in the ground. During a visit in 2018, men leaned on their *bordunas* and held forth in a style best described as livid tirades, drilling the sticks into the floor for emphasis. Tamasaimpa was incongruously at ease through these vehement monologues, because, it turned out, the Korubo were remarking on mundane matters like pig hunting and gathering *buriti* palm fruits, as well as more serious concerns, like malaria and armed fishermen invading their territory. Their gestures, postures, tone of voice, expository style, and physical contact were all utterly their own, unalloyed by the norms of the big culture outside.

Megaforests have nurtured thousands of ways of knowing and describing nature—and the experience of living generally. Megaforest people still know how to heal with boreal tree bark, share a jungle with elephants, move in their environments with supreme competence, and talk about them with a precision that is a gift to modern science. And, in secure megaforests, people can carry on relationships with their surroundings that are old and familial. Tamasaimpa's cousin, Kenampa, coordinator of the Union of Indigenous Peoples of the Javari Valley, puts it like this: "The forest is part of our family. When we look at the forest, we don't just see forest. We see lives. Lives that need us just like we need them."

6

Guardians

A carbon atom in the Amazon is thirty-six times less likely to be vaporized if it's in an Indigenous-controlled forest than if it happens to reside on unprotected land. Even compared with protected areas, Indigenous forest carbon is more secure, by a factor of six. Carbon loss on unprotected lands is due mostly to outright deforestation—ecosystems cut and usually burned to bare ground. By contrast, 82 percent of the carbon lost from Indigenous lands is emitted by partial forest disturbance from selective logging and garden clearings, which tend to grow back, as well as to more significant and unavoidable problems that originate elsewhere, like wildfires and climate change.

Our Marubo friend Tamasaimpa says this shouldn't surprise anyone: "We have people living in there. If you don't have people there, if you only have deer and agouti and little birds, what logger or miner is going to obey the deer and agouti and birds? If people want to come in and take our forests, they have to kill us."

Forest people are often willing to risk it all for their land. Like one time when Tamasaimpa took on a detachment of heavily armed Peruvian National Police. That was in 2000, when he was 25. He was working on a German-funded project to demarcate the Javari Valley Indigenous

Territory's boundary and protect it from illegal resource extraction. Similar work was afoot all over the Amazon as part of the G-7 countries' Pilot Program to Conserve the Brazilian Rainforest, which originated from the 1990 summit hosted by President George H. W. Bush in Houston.

Before the 1990s, warfare between tribes had been common. Conflicts were resolved with guns, arrows, clubs, and spears. Men kidnapped women from rival tribes. Notwithstanding the ethnic mixing this involved, cultural identity was—and continues to be—extremely powerful. As a youth, Tamasaimpa viewed the neighboring Matsés,* who occupy an area straddling the Peruvian border, as his people's archenemies. Tamasaimpa's aunt was carried off by the Matsés, who suffered bloody retribution at the hands of his father.

By the late 1990s, two things had changed in Brazil. Government agents from FUNAI were increasingly present in the Javari Valley, and they discouraged tribal warfare. And the various tribes realized they had common enemies closing in. Petrobras, the national petroleum company, had drilled a few exploratory wells. Hunters came for wild pigs, deer, tapirs, and especially for the turtles that lay eggs on the river beaches during low water. Fishing gangs came for the canoe-sized and delicious arapaima to supply distant urban markets. And loggers were sacking the forest's valuable stands of timber. The diverse peoples of the newly created Indigenous reserve started to band together in a Javari pan-ethnic movement comprised of five culturally and linguistically distinct contacted groups.

The Matsés in particular were having trouble with Peruvian loggers, who crossed the Jaquirana River into Brazilian territory and illegally hauled out vast quantities of mahogany with a wink and a nod from Peru's police and environmental agencies. The Jaquirana squiggles its way through the jungle, Peru on the left and Brazil on the right. It flows into the equally serpentine Javari, which gives way to the Solimões hundreds of miles downstream. Loggers were floating rafts containing hundreds of logs down the river, past both Peruvian and Brazilian authorities, using

* Matsés is the term used by the people themselves and generally in Peru. In Brazil the Matsés are also called Mayoruna.

fake papers that indicated Peruvian provenance. Tamasaimpa alerted the Brazilian Federal Police, the army, and the national environmental regulatory agency, IBAMA, that the Peruvians were stealing. The authorities wanted proof. Tamasaimpa got the Matsés to mark the logs leaving their territory. Officials still didn't act.

In Marubo and Matsés cultures, leaving the loggers' territorial violation unanswered was problematic. It damaged their forest and showed weakness, which, in their long experience, would encourage more trouble later. Tamasaimpa knew the ferocity of the Matsés from their long and violent entanglement with his own family. So, with $20,000 of project funds, he stocked up on gasoline, food, guns, and ammunition. He convened 200 Matsés men, who painted themselves for war, organized into three groups, and seized the next three Peruvian barges, loaded with up to 700 stolen logs.

"The Matsés were all in. They were so enthusiastic. They tied up the crew," recalled Tamasaimpa one evening in late 2019 over dinner in the town of Tabatinga, just outside the Javari territory. "All I did was help them 'let their dogs out!'" Peruvian police were dispatched to rescue their captive countrymen. Word of their approach traveled by the village radio grapevine. Tamasaimpa and the Matsés discussed the situation. "That's when we decided to ambush them."

At around 2 o'clock in the afternoon the next day, he stood on a log barge with five bare-chested Matsés painted red and black, in various patterns according to their clan affiliations: red dots, black hands, red masks, black necks, and various configurations of stripes, along with *pataúá* palm headbands decorated red with *urucum* dye. This is how the Matsés dress for trouble. These particular fearless Matsés were handpicked older warriors whose best fighting days were behind them: the bait.

"The police showed up armed to the teeth," Tamasaimpa said, remembering the Peruvian commander's threat. "He said to us, 'If anything happens to you here, no one will ever know.' They pointed the guns right in our faces. I said, 'Go ahead and kill us but you are also going to die.'" Then Tamasaimpa whistled.

Twenty armed Matsés warriors appeared from behind the bushes and

Tamasaimpa (left) and Matsés leader Waki Mayoruna, years after their log barge operation.

trees on one sloping riverbank. He whistled again and a similar number emerged on the other bank. They brandished their new firearms and chanted. Tamasaimpa pushed back from our dinner table to demonstrate, lowering his torso, holding his arms out slightly as if to grapple, and producing a throaty chant with a mean look on his face. Then he cracked up.

A standoff ensued and spiraled into a diplomatic brouhaha. Elite special forces hardened in Rio de Janeiro favelas were dispatched to the scene. The two countries' ambassadors flew into the remote jungle and, over the next few days, gradually de-escalated the situation. The Peruvians never did get the mahogany back.

Tamasaimpa loves this story, in which the Matsés go to war for their rights. He revels in the flat-footed astonishment of all the outsiders, who were accustomed to having their way with the tribes. Governments, even those with good policies on Indigenous land rights, often display a chronic indifference to enforcement, which throws territorial defense back on Indigenous peoples. "The Matsés didn't know they had power, but it was inside them all the time. They weren't afraid to die. We didn't have trouble with the loggers after that."

Article 231 of Brazil's 1988 constitution, which established the country's post-dictatorship civic order, reads as follows:

> Indians shall have their social organization, customs, languages, beliefs and traditions recognized, as well as their original rights to the lands they traditionally occupy, it being incumbent upon the Union to demarcate them, protect and ensure respect for all of their property.

It's hard to overstate the impact of this short paragraph. The "lands they traditionally occupy" denotes the places they historically roamed and foraged, not just the villages they've been stuck in since the arrival of missionaries. It was a revolutionary upgrade in Indigenous territorial control. The government had five years to demarcate all the Indigenous lands.

Upper Rio Negro Indigenous Territory, Brazil.

Even though the job is only partially complete over three decades later, nearly 300 million acres (three Californias) of territory have been officially recognized in the Amazon. While the lands are formally held by the federal government and considered units of Brazil's protected area system, the Indigenous residents have permanent, exclusive use rights, held collectively, which means they can't sell their lands and can't parcel them out into individually owned plots.

Nearly all Amazon countries recognize Indigenous territory in one form or another. None, however, has a policy as robust as Colombia's, which establishes inalienable territories called *resguardos*. Indigenous Colombians own 77 million acres of *resguardos*, a full quarter of the country's territory. Virtually all of it is covered by intact forest. Good law doesn't solve everything; there's plenty of pressure on Indigenous forests, even in the cases of the basin's land-rights leaders, Brazil and Colombia. Nonetheless, their policies exemplify three essentials of Indigenous territorial control: exclusive rights, full traditional territories, and inalienability.

Globally, Indigenous communities have some measure of legal control on 36 percent of the IFLs. That's nearly a billion acres of forest. All five megaforests have, to greater and lesser degrees, populations of their original peoples. Any credible plan to save the planet's intact forests, therefore, necessarily involves supporting Indigenous forest peoples' claims to and influence over megaforest territories.

That is not to suggest that all Indigenous peoples are unswerving preservationists whose lands are besieged by crafty outsiders. The narrative of noble Indigenous victims is a demeaning caricature. Indigenous communities usually have mixed and nuanced views about resource extraction. Some favor tapping timber, minerals, and other industrial resources where it's possible without doing irrevocable injury to the land. Others prefer to stick with traditional livelihoods. These are active debates in communities worldwide. Overall, however, in traditional forest communities, money's allure is more likely to operate within guardrails of their practical, day-to-day dependence on, and kinship with, the woods.

Across Canada, First Nations are playing more central roles in forest protection. That's thanks in part to Valérie Courtois, a forester and member of the Innu First Nation in Quebec and Labrador. Confident and articulate, Courtois founded the Indigenous Leadership Initiative (ILI) in 2013 to foment a national movement of Indigenous "guardians." The guardians care for their traditional territories, which are sometimes formally recognized and sometimes not.

Courtois says that the movement began with a canoe trip made by a young Haida man known as Captain Gold in the early 1970s. Fifty years earlier, the 1918 global flu pandemic nearly wiped out Captain Gold's people, the Haida, who had lived for over 10,000 years on a claw-shaped archipelago called Haida Gwaii off the British Columbia coast. The remnant population was settled in two mainland villages. That left their island rainforests temporarily vacant.

"Captain Gold had a dream to go revisit where he was from. It's a village site call Sgang Gwayy. So, he ordered a canoe from Sears and Roebuck—at the time he had never canoed before—and he made a makeshift sail," Courtois says. Around 30 at the time, Captain Gold paddled 150 miles, crossing the often-tempestuous Hecate Strait, and eased the department-store vessel, stern first, as is Haida custom, onto the shores of his family's land. He started cleaning the place up. Others followed his example. Soon the southern part of the Haida Gwaii archipelago was a setting for cultural recovery and a bit of tourism. That was the start of the Haida Watchmen, the first guardians. They have watched and defended territory and served as guides for visitors. Working off this template, Courtois's Innu people, a caribou-hunting First Nation on the other side of Canada, established guardians in the early 1990s. Courtois ran the program from 2003 to 2009. Guardians sprang up with increasing frequency across Canada and, as of early 2021, numbered seventy groups and counting.

Asked if there are scuffles between guardians and violators, she says very few: "Eighty percent of enforcement is presence. There's something to be said about behaviors of people when they're being watched. It changes, and it tends to be much better." Courtois is the daughter of a policeman but thinks unarmed guardians work better than conventional law enforcement for traditional lands. "Fundamentally, people who are hunters, who are trappers, who go to the land, they do it for all kinds of reasons but generally speaking there's a commonality of love for the land or appreciation for nature and that's a good place to start from for those discussions of behavior."

Tanya Ball is one of the young leaders spreading the guardian model in remote parts of the boreal forest. She leads the program in the southern part of the Kaska territory, which lies within British Columbia, immediately south of the Yukon. Ball tried social work and culinary school before discovering that her real calling was to be out in the mountains and forests.

She hunts and fishes and, when we talked in 2020, was eagerly looking forward to taking her first moose, a major milestone in her culture.

Ball's group first organized in 2015 in response to problems during the annual fall hunting season: "We were noticing an influx of people coming in, overcrowding, a lot of environmental concerns and we were getting a lot of complaints back from our people." She and a tiny handful of fellow Kaska members started going out on the land, noting where hunters were and talking to those willing to engage. Many were not. "I think they have the stereotype that First Nations don't want people in our territories, and they've got a misconception of what the guardian program is."

For Ball, Courtois, and their fellow guardians, the program is not an effort to forcibly wrest resources away from non-Indigenous people. It's about being *present*, acquainting youth with the source of their people's identity, walking old trails that predate roads, planes, and snowmobiles, walking there with elders who can tell the stories as they go, and layering science with culture. Ball's guardians gather water quality measurements, climate data, and wildlife sightings, making iPad entries on forms that use Kaska names for all the animals. Her biggest smile of our interview came when she talked about starting multiday patrols: "Just getting back out on the land. I think it's vital for our young people to start learning that and going back out, having that connection. I'm really looking forward to that! I get to go camping for work!"

The program now operates year-round, collaborates with BC government conservation agents, and is linked to the guardians of the neighboring Tahltan and Taku River Tlingit peoples. Ball coordinates this three-way association. In her ideal future, the territory buzzes with the energy of its people, out on reopened ancient trails, sleeping in camps and cabins, youth and elders together in the woods and on the rivers. Talking about this animal, that tree, lichen, or fish, and calling them all by their Kaska names.

The guardians' leverage to assert themselves in the Canadian megaforest has roots in a policy decision made over 250 years ago, buttressed by a twenty-first-century Supreme Court ruling. The North American colonies, along with subsequent nations of Canada and the United States, seized Native lands with the punitive zeal of war victors and the condescension of "civilizing" powers. With few exceptions, they relegated Indigenous survivors to tiny territories, often areas with which tribes had no historical relationship. These arrangements were formalized in treaties between the Native peoples and the newcomer governments. But the remote north and west of Canada, as well as Alaska—a huge portion of the North American megaforest—remained treatyless into the 1970s. At that point Indigenous groups occupying some of North America's best-preserved woodlands started to realize that their land had never been officially given up—only unofficially taken.

In the 1980s and 1990s, the Haida Watchmen and others witnessed and protested as loggers chewed their way through cedar and hemlock titans on Haida Gwaii. Companies extracted 21 billion board feet (50 million cubic meters) from 420,000 acres, reaping $20 billion in revenue. The provincial government's cut was 10 percent, so it readily issued harvest permits. British Columbia denied the Haida's existence as a distinct culture, notwithstanding their international fame as carvers of monumental totem poles. Even if the Haida were shown to exist, both the province and Canada denied that their living members had rights to the trees or anything else in their homeland of 500 generations.

In 2004, the Canadian Supreme Court surprised the Haida and everyone else by unanimously ruling that the First Nation did exist and had never formally yielded their land. This judgment was based on a 1763 proclamation issued by George III of England, which has come to be known as the Indigenous Magna Carta. In it, the king said that until the Crown bought land from its Aboriginal occupants, those peoples owned it. He may have had something else in mind—perhaps a monopoly on the real estate market in North America—but the consequence, the Court held, was that Indigenous people still legally owned lands not covered by treaties signed after 1763. Because the titular ruler of Canada was Elizabeth II, George's great-great-great granddaughter, large expanses of land in the Canadian West and North, especially in British Columbia, were unceded territories still belonging to Indigenous peoples.

The Haida's Supreme Court win did not suddenly rewind two centuries of history and return the land, waters, and resources to First Nations. But it did improve their bargaining power in ongoing and new treaty negotiations.

One such treaty is the Umbrella Final Agreement between the Canadian government, Yukon Territory, and eleven First Nations. The two Kaska First Nations in the Yukon (there are three adjacent Kaska groups across the British Columbia border), Ross River and Liard, initially joined twelve other Yukon peoples in the talks. Strength in numbers was the idea. They all sketched out their ancestral territories on a map, resolving areas of overlap. The final deal recognized Indigenous ownership of an average of 8 percent of their traditional territories. On half of that 8 percent, First Nations would own minerals and hydrocarbons under the ground, with rights limited to surface resources on the other half. The Yukon government would retain the remaining 92 percent, "comanaging" it with the Indigenous peoples of each particular area.

First Nations started signing the deal in 1993. Both of Yukon's Kaska communities and one other First Nation walked away. There was nothing in their past interactions with colonial society to convince the Kaska that comanaging 92 percent of their land would go well. Instead, the Kaska of the Yukon decided to assert their traditional ownership and fight in court. They won a 2012 moratorium on new mining claims in their territory and continue to discuss their land rights with Yukon's government. In 2018, the Ross River Kaska started issuing their own hunting licenses, which come with education about sensitive areas of the territory. These are a voluntary addition to those required by the Yukon government. In their first year, around a quarter of hunters obtained the Kaska permits. As of 2021, the Ross River Kaska are pressing ahead with a proposed Indigenous protected area that would infuse their values and environmental concerns into management of the region.

The Heiltsuk, who have inhabited the British Columbia coast north of Vancouver Island for millennia, also took part in the modern treaty process for a time and then realized they were better off simply stating what is theirs: 8.6 million acres of forests, fjords, lakes, rivers, and islands teeming with grizzlies, salmon, orcas, whales, and eagles. On a video call from her office in Bella Bella, British Columbia, Heiltsuk leader Jess Housty said, "It became apparent that there was so much more for us to lose than we possibly had to gain through that process, and we pulled out." The First Nation gave up the legal certainty of title in order to assert their own understanding of their place in the landscape. That vision is enshrined in a unilateral Heiltsuk Declaration of Title and Rights signed by the Heiltsuk's two governing bodies, the traditional *Hemas* hereditary leaders and the elected Tribal Council.

The First Nation built cabins where they had traditionally fished, hunted, and collected food. De facto occupation of lands provided influence for the Heiltsuk in a multiparty negotiation over a vast region called the Great Bear Rainforest that encompasses their lands. A 2006 agreement set aside sizable areas for preservation. The Heiltsuk allow logging but require that companies use enlightened forestry practices known as ecosystem-based management.

Jess Housty's obligation to guard the Heiltsuk territory is written on the birth certificate of her first-born son, Noen. It's a contraction of "No Enbridge," referring to a firm that proposed installing a pipeline across Heiltsuk and several other First Nation territories. The Heiltsuk successfully fought the project. At a public hearing in 2012, Jess's brother William gave a dazzling 45-minute seminar on the obligations that connect the Heiltsuk's culture and language to every aspect of their ecosystem—shellfish, salmon, trees, and even the salt water, which they imbibe for spiritual purposes. The lecture detailed the physical and ethereal attachments that define the people and the sensitivity of those connections to events like crude oil spilling into the coastal waters. The pipeline was defeated.

Legalizing ownership, Jess Housty said, risks sapping the power of more important principles: "People talk a lot about asserting rights and asserting title and sovereignty and, I mean, I guess you can think about it that way but for me it's about asserting relationship. We occupied a very large territory that has lots of different kinds of geography in it. We were tied to those places through our relationships to them. If you look at translations of different Heiltsuk place names throughout the territory, often the names refer to our relationship there, to the things we harvested, events in our history that happened there, people who came from there. And when you look at our names as Heiltsuk people, the translations really relate to how we interacted with our landscapes as well. The thing I struggle with when people reduce it to a question of rights and title is that it starts to feel like it's about ownership and I don't feel like that is the point. It was always about your reciprocal obligations to place, the way places took care of you and you took care of them in turn and helped each other to thrive. I think that happens through very careful relationships with the places we live."

The western boreal and temperate rainforests extend right into Alaska, but thanks to the violent nature of America's separation from the British Crown, King George III's 1763 proclamation stops at the border. The status of Alaskan Native lands remained in limbo 100 years after the United States' purchase of the territory and over a decade after its 1959 statehood. In 1971 this all changed. Congress passed the Alaska Native Claims Settlement Act (ANCSA), turning Native communities into companies called Alaska Native Corporations (ANC). Twelve regional corporations were set up as owners of 44 million acres—10 percent of the state, including the subsurface resources—and received $1 billion in compensation for the 90 percent of land that went to state and federal government. That's $2.63 per acre. Over 200 village corporations overlapped the twelve regional companies, with rights to aboveground resources like timber.

Settling Indigenous land claims had been in the works for years in order to open Alaska for development schemes. One idea was to cork the Yukon River with the Rampart Dam to make the world's largest artificial lake, with a power plant big enough to supply everyone in Alaska at least twenty times over. Another, Project Chariot, was to obliterate the coastal Inupiak village of Point Hope with five nuclear bombs and sail ships into the resulting harbor. Neither came to fruition, but the Prudhoe Bay oil strike did and required a pipeline across Native lands. Congress fast-tracked the ANCSA and the creation of the Native corporations.

As claims settlement approached, the Alaska Federation of Natives formed and was in the thick of negotiations for the corporations, some of which continue to make billions from oil, gas, coal, logging, and an array of lucrative contracting ventures. The companies are loci of economic and political power in the state and, some advocates note, share revenues in ways that echo Native customs.

But the Alaska Native Corporation is a uniquely awkward model of social organization: community as business. We're unaware of any other human community, Indigenous or otherwise, that is legally structured as a for-profit enterprise. In the American lower forty-eight and Canada, tribes and First Nations own businesses collectively. In Alaska, the business, in a sense, owns the community. It holds title to all their land. The Native people hold individual shares in the corporation, rather than owning it as a group. Indigenous land control was built to dissolve, piece by piece.

A dozen years after the act was passed, the Inuit Circumpolar Conference asked Thomas Berger, a former British Columbia Supreme Court justice, to study its impacts. He traveled to sixty-two villages and listened. His report is scathing. Berger details the risk of Natives losing lands in fire sales to pay off business losses, legal fees, and taxes, noting the lack of substantial benefit to the community members from the mostly failing enterprises and highlighting the affront that industrial ventures present to subsistence traditions of the Natives.

The power of the report, however, is not in what its author says. It's in the hundreds of direct quotes from Natives. In a meeting in the town of Kotzebue, Native Pete Shaeffer told Berger: "God chose to put a wealth of natural materials in our land and, like a magnet, the wealth attracts those who have no honor or respect for the people who live on the land. Systematically, they conspire the easiest way to get the land. What they dreamed up was a plot for that purpose and, amongst it, was the Alaska Native Claims Settlement Act."

Of the three criteria for strong Indigenous land control, the law meets the standard of exclusive use, but it covers only a fraction of traditional territories and makes the Native lands eminently alienable. The act was amended after the Berger report to reduce chances of land losses, and so far, the corporations have held on to territory. However, Steve Kallick, of the International Boreal Forest Campaign, says, "Lands have been estranged from the tribal governance entity." Kallick, who spent over ten years in the state advocating for forests, adds, "The corporations were created to assimilate Indigenous peoples and make profit, not to steward the lands." Unfortunately, even profit has eluded many of the corporations and, in the more successful ones, failed to trickle down to their shareholders. "Any village you go to and ask, 'what have you gotten out of the Native Claims Settlement Act?' 90 percent would say, 'very little.'"

It would take a heroic effort to unwind the debacle of the Alaska Native Claims Settlement Act. Rather than redraw entrenched maps or reform institutions, Kallick sees a different solution for the US boreal forests: Canadian-style Indigenous guardians. To confront problems such as poaching, fire, and mining, Kallick says Indigenous villages and others could design cross-ownership land-use plans and put Native guardian boots in the boreal forest. He's working on bringing Canadian guardians from Courtois's network across the border to share experiences with Alaskan communities.

The colonization of Alaska that culminated in ANCSA started with Vitus Bering's 1741 reconnaissance voyage sponsored by the Russian empress Anna Ivanovna. It was a late chapter in the eastward expansion project that began in the shadow of the Ural Mountains. Cossack fighters had bested Turkic Central Asian rulers just east of the Urals in the early 1600s, clearing the way for colonization of Siberia. Within a few decades Russian forts were set up all the way to the Pacific with officials collecting fur tribute from Native peoples, stealing wives and children and killing the locals. An estimated 80 percent of Evenks and Sakha perished in the first smallpox wave. Sexually transmitted diseases and measles followed. European homesteaders came along behind the guns-germs-and-fur tax vanguard.

The 1916 completion of the Trans-Siberian Railway cemented the integration of the Indigenous homelands into Russia. After a brief postrevolutionary celebration of Indigenous rights, Stalin's Soviet system reversed course. In an all-out assault on Indigenous identity, the state appropriated land, persecuted ethnic minorities, suppressed Buddhism and shamanism, and outlawed Native languages. When the USSR collapsed, there was another wisp of hope for Indigenous rights, but it has proved fleeting.

Thirty years after the dissolution of the Soviet Union, there are big forest regions under Indigenous governance—on paper. The Komi, for example, have their own republic, one of twenty-two established for ethnic minorities. It's an area the size of France nestled along the western slope of the Ural Mountains and contains the largest expanse of forest in European Russia. In practice, explains Svetlana Turubanova, the University of Maryland forest mapping expert, her fellow Komi people have little say about how the forests of spruce and fir are used. Moscow decides. It's the same in Buryatia, where massive forests are richly entwined with the Buryat culture. In practice, the federal government owns the forests of Buryatia and that of all the other republics.

Ethnicities with fewer than 50,000 members can petition for a Territory of Traditional Nature Use (TTNU) on their ancestral lands. To learn more, in May 2019 we drove all day with the Tunkinsky tourism official Sergey Natvevich to a remote valley in the heart of the Sayan Mountains, where we hoped to talk to a Buddhist lama called Norbu who set up a TTNU for his own Soyot people. From Arshan, we drove west until the Tunka Valley funneled to a point and the Irkut River carved into low hills. To increase our chances of safe passage, we offered rice, tobacco, and coins to forest spirits at a roadside birch grove bedecked with prayer ribbons.

Very soon, we needed the help. As we neared Mongolia, we came to a border zone checkpoint. The area close to the border is tightly restricted. Normal procedure is to request a permit several months in advance, and foreigners without permits have been known to get thrown in jail. But we decided to visit on the spur of the moment and lacked permits. At the checkpoint, Natvevich held a lengthy discussion with the officials about our mission. Our passports disappeared for a long time. Phone calls were apparently made to Moscow. An official came out of the guardhouse and asked Natvevich questions while gripping the stack of passports. The man wanted to see if we had any photos of the spiritual leader we claimed to be visiting. As it happened, John had a shot on his phone with a robed Norbu Lama from a recent meeting in Mongolia where the two were guests at a spring ceremony in which shamans summon migratory birds back to the steppe. In the picture, the two have hands together in prayer position.

"Ah, you are Buddhists!" the guard chuckled. Convinced we were harmless, he waved us through.

Another 5 hours on a dirt road brought us to Norbu's home town of Orlick, in the very heart of an intact forest woven around mountain peaks. In the burgundy robes and headgear of the Tibetan Red Hat school, Norbu described his youth in the alpine forests. He said with a grin that he has always been a crack shot. He started hunting in the woods so young that the rifle butt dragged on the ground when he slung the weapon over his shoulder. He was later a sharpshooter in the Soviet army. After the Soviet collapse, Buddhism became legal again and he embarked on the spiritual life. He gave us a tour of the paintings of various spiritual protectors in his mountain temple. One arresting image shows a horse with an eye on its rump to monitor enemies approaching from behind.

His Soyot ancestors have been keeping an eye on these mountain forests as far back as anyone can remember. Norbu Lama spearheaded creation of a Soyot TTNU to give locals more control over timber, hunting, yak grazing, reindeer herding, and other resources that sustain them. Indigenous peoples have created hundreds of TTNUs under regional guidelines, with the aspiration of managing, or comanaging, traditional territories. In a few places they have been able to veto some industrial development. Meanwhile, Russian law has steadily chipped away at the promise of the TTNUs. Even before the ink was dry on the law that enabled the TTNUs, a separate lands law was enacted that eliminates the possibilities of communal land control essential to some traditional lifestyles.

We asked Norbu what practical difference the TTNU could make for the Soyot people. He explained that while the 6.4-million-acre TTNU is legally recognized by the government of the Russian Republic of Buryatia, there's no official guidance on how to implement the special territory. At the time we met, for instance, he was trying to legalize traditional use of reindeer pastures. The animals were first domesticated right here in the Sayan mountains, so nothing could be more traditional. And yet there were conflicts to unravel with the national government, which owns 90

Norbu Lama promotes Soyot land use traditions in the Sayan Mountains of southern Siberia.

percent of the land. Like nearly all of the 500-plus TTNUs created so far, the Soyot area is recognized by the regional government but not by federal authorities.

Norbu deftly steered the conversation away from the TTNU politics and back to the temple walls, continuing our clockwise tour. The minutely explained images started to feel like a painted koan—featuring guardians, demons, heroes, wise men, celestial battles, and dynastic spans of time—directed at a questioner wanting simple answers about how to save the forest.

Pavel Sulyandziga is an Udege Indigenous activist from the Taiga's biodiversity hotspot in the Russian Far East. The Udege share territory with Amur tigers, whom the people respect as descendants from a common ancestor. *Udege* in the Udege language means "forest people." Sulyandziga is also the former head of an alliance of the sub-50,000-member Native groups called the Indigenous Small-Numbered Peoples of the North, Siberia, and the Far East. As a national leader, he helped secure passage of the 2001 TTNU law, which he hoped to apply in his own people's forest in the Bikin River valley. Loggers had already cut their way through the lower Bikin and, in doing so, reduced Udege cultural subgroups from eight to four. The rest proved resilient. Sulyandziga, a mathematician, explains that his grandparents' generation exhorted the community to pursue modern education but resist outside religion. Education gave them articulate advocates to confront adversaries like the Hyundai Corporation, which had already clear-cut other forests in the Far East. In the early 1990s, Hyundai received government permission to log still-intact portions of the Udege's land. The Udege stood firm. International criticism deluged the Korean giant, which eventually withdrew, sending a vice president to personally apologize for having had the temerity to think about logging the forest people's forest.

But Sulyandziga says his attempt to set up a TTNU met local opposition, including from some Udege, whom he says had been co-opted by outside businessmen. Then the Udege's tiger-kin brought a different opportunity to save the forest. "I don't like Putin," said Sulyandziga, who now lives in exile in New York, by video call, "but Putin likes tigers and for that I thank him." In 2010, Russia had hosted the first major gathering of the Global Tiger Initiative, a conservation effort that included thirteen tiger-range countries, the World Bank, and the Smithsonian. Russia proposed to protect its tigers by creating a national park in the Udege's forests.

The Udege said no.

While they revere tigers and observe a taboo against killing them, the people did not want to relinquish control to federal park managers directed from Moscow who might cut off access to the flora and fauna that sustain villagers and their culture.

But the Udege left the door open to a deal. They made seven demands, two that required legislative changes, including a change in the stated purpose of protected areas. "Before, the national park was for protecting nature. Now it's for protecting nature and the development of Indigenous people," Sulyandziga explained. The Udege insisted that the park rules not encroach on their customary ways of using nature, including hunting. Further, they demanded that an Udege leader be either director or vice director of the park. To their astonishment, Putin agreed. The Bikin National Park debuted in 2015. Sulyandziga reports that it's been a big success for his people. Apart from conservation of the 2.7-million-acre forest, their hunting and fishing rights, and control over park management, they got electricity, Internet, a health center, and cash income for hunters who know the territory intimately and work as guards. "We have a soccer field, very good quality. Bigger cities do not have fields like we do, deep in the forest," Sulyandziga laughed.

Amur tiger populations had rebounded by 50 percent between the late 1990s and the designation of the park. They have remained stable at around 500 to 600 animals into the early 2020s thanks to protection within Udege territory and other conservation efforts. The park, rather than attenuating their access to resources, gave the Udege the kind of inalienable territorial control that has been difficult for communities to obtain through direct recognition of their land rights in the Taiga.

The only megaforest where Indigenous land rights are more tenuous than in Russia is in the Congo. The territories where Pygmies traditionally hunt, collect plants, migrate, and worship are occupied by either strictly protected parks or heedlessly exploited by logging and mining companies. Entrenched subservience to Bantu "symbionts" complicates the notion of Indigenous territorial recognition; acknowledging Pygmy lands without reference to their longtime social partners could be viewed as an attack on those ties. As such there is little formal recognition of the Pygmies' unique connection to forests.

One day as we drove through the Congolaise Industrielle des Bois logging concession near the Nouabalé-Ndoki National Park, we saw the smallest of nods to their status in the form of signs designating hunting grounds for the local people. We were riding with Gaston Abea, the Aka researcher who studies the truffle-eating gorillas. He scrutinized the logging roads that pierced the forest at intervals of every few hundred yards. Fresh sapele logs lay alongside the road, ready to be trucked to the mill. An old Aka man sat under a tarp on a dirty mattress next to a tiny campfire. "They are going back in to cut again in places they have already been," Abea lamented.

Non-Pygmies' long-standing Congo presence, the intimacy in which the two cultures live, and persistence of phenotypic and cultural differences complicate the notion of "Indigenous" in the Central African megaforest. Perhaps because the Pygmies are most connected to nature and are socially disadvantaged, the Indigenous label—*autochthones*—is applied exclusively to them by the forest region's governments, all of which signed the 2007 United Nations Declaration on the Rights of Indigenous Peoples, known as UNDRIP. The declaration commits adherents to recognize Indigenous territories. The Republic of the Congo even has its own 2011 legislation vouchsafing land to its autochthones. So far these are paper pledges, according to Victoria Tauli-Corpuz, the UN Special Rapporteur for Indigenous Issues from 2014 to 2020 and a member of the Kankanaey-Igorot mountain tribe in the Philippines. She visited the country in her official capacity in 2019, finding that Pygmies' ancestral forests are subsumed within logging leases or protected areas. Tauli-Corpuz underscores "fortress conservation"—locking locals out of their longtime territories—as a particular problem in the Congo. In a setting where there is no official recognition of forest peoples' territories, protected areas are drawn without the sort of rights the Udege secured to keep using their ancestral forest.

Tauli-Corpuz has a disarmingly friendly way of conveying moral outrage. "The way they talk as government officials about Indigenous peoples is so unacceptable. I cannot forget this President of the Congress whom I met with. I told him what I saw in the country and I asked him, 'How many members of parliament are Indigenous peoples?' He said, 'None.' I said, 'Not even one?' He said, 'No, but my driver is an Indigenous person.'" Tauli-Corpuz laughed heartily at the absurd, racist remark. "When that kind of culture exists, even if you have a good law, I don't see anything positive in the future as far as forests and Indigenous peoples are concerned."

The newest potential Indigenous rights law in the Congo megaforest was introduced in the DRC's National Assembly in June 2020. If the cultural obstacles of the kind Tauli-Corpuz saw in neighboring Republic of the Congo don't get in the way, this law would provide the DRC's Indigenous Pygmies with rights like those in policy exemplars such as Brazil. In 2014, the DRC passed another law, which, while not Pygmy-specific, enables forest peoples to set up 125,000-acre community concessions. For this model to save forests and benefit the villagers, locals must be well organized and draw up technical management plans that pass government muster. And they have to find a way to keep the village elites or militias from commandeering the concessions.

It sounds like a tall order, but Dominique Bikaba thinks it can work. He's the founder and executive director of the Strong Roots organization, which helps communities establish concessions. Bikaba grew up in a forest that's now a national park called Kahuzi-Biega, in the eastern DRC. Although his family was kicked out, Bikaba is remarkably supportive of the park. He visits regularly to spend time with the eastern lowland gorillas. They inspire him to keep striving for a balance between people and nature in one of the most challenging conservation settings on Earth.

His main focus is a Yosemite-sized forest between Kahuzi-Biega and the Itombwe Nature Reserve. There he's working to set up community forests under the 2014 law with seven groups, which include both Bantu and Pygmies. "Everyone's engaged, because once this is declared, the government cannot give a title for mineral exploitation or create a national park or something," he said. In an auspicious sign for the community forests, villagers in the area have respected a ban on gorilla hunting imposed by a local tribal king in 2010. Bikaba said that since then, gorillas both outside and within the protected areas have been thriving. "Communities are not as complicated as we think they are. If they understand, they engage, if they trust, they go along. . . . They don't need a million US dollars. They can make a big change with nothing."

New Guinea is at the other end of the Indigenous territorial control spectrum from the Congo. Native people hold over 90 percent of the island. Just as Heiltsuk leader Jess Housty expressed, ownership in New Guinea is a crude reduction of people's land relationships. Customary claims can be fluid, dynamic, overlapping, complex, and debatable in ways that permit adaptation to changing needs within communities. A modern ownership claim can be both a defense against having land stolen and the first step toward selling it forever.

As we walked in her clan's forest, Fince Momo stopped and pointed out a huge strangler fig with a broomlike base of narrow root-trunks. We didn't notice the moss-rimed bones at first. Two dozen wild pig mandibles showed their teeth from a tether strung among the exposed roots. This was a *sruon*, Momo explained, a sort of ancestor neighborhood, where the spirits of departed clan members abide. It is hard to imagine the Momos selling off their ancestors, sacred mountains, a crystalline river, creeks, waterfalls, millions of trees, and birds of paradise. Momo vows that they won't, no matter how big a mountain of gold they might be sitting on.

Other clans do sell. Yunus Yumte, of the Samdhana Institute, says that in Sorong, the regency that adjoins Tambrauw to the southwest and holds the province's biggest city, the demand for property is high and forests are falling. Indigenous Papuans had sold 57 percent of the land in Sorong by 2020. The Samdhana Institute is working to avert the repetition of Sorong's story by securing customary land rights.

As a first step, they sometimes map clan boundaries. The idea is to ensure that as developers arrive, clans can show what is theirs and avert conflicts with neighboring clans. This is a complicated business. There are always overlapping claims. Resolution requires lengthy, mediated historical discussions to establish which group has the senior claim to each area. Samdhana and partner groups have mapped territories of several clans per year. They have a long way to go; in Papua and West Papua overall, there are more than 250 tribes. The Abun tribe alone, Tambrauw's largest, has thirty-two landowning clans living in eighty-five different villages.

Putting aside the practical reality that a master clan map could take decades to prepare, questions arise as to whether or not it's a good idea. The social topography of land rights is every bit as convoluted and daunting as the abrupt slopes and creases of physical geography. "Everyone is scared of Papua because our tenure situation is so confusing!" Yumte says of prospective investors. "Once you clarify them, the maps can easily go to other hands so the land can go to other hands." In fact, he says, companies themselves are also working on customary boundary maps with communities in hopes of accelerating access to resources.

That brings us to the Samdhana Institute's second prescription, customary governance, in which tribes take a stronger role in discussions over what resources clans decide to sell. Abun leader Kundrat Yeudi is trying to establish customary government for his tribe. His own clan's lowland forest was logged from 2007 to 2013. A few elders got paid, deer and wallabies became scarce, and erosion clouded creeks. Yeudi was a teenager at the time. We talked to him in Abun territory in West Papua in early 2020. At 27, the bearded and soft-spoken Yeudi now had five children and gravitas. He had convinced some villages to rebuff loggers and was leading the Abun tribe in an effort to respond collectively to offers of money for the things in their ancestral forests. "Our tribe needs a traditional institution because economic value changes our perception of land," Yeudi said.

The third leg of Samdhana's strategy is to find other ways for people to make money. Taking tourists to the woods to see birds of paradise and learn about the forest cultures is one alternative. As we sat on a small beach on the Sumi creek, Yumte pointed to tubular rattan stems on the opposite bank as another option: furniture. Yumte is under no illusions that forest-friendly microbusinesses are going to churn out the billions that Freeport-MacMoRan makes digging up gold and copper. But he's hopeful that smaller sums will suffice for clans like the Momo and the Yeudi, who already get most of what they need from the forest.

The final part of the strategy involves building character among Papua's youth. Until around 2000, selected adolescents received traditional education all across the province. The instruction, delivered in the forest, is called *wuon* for boys and *fenya meroh* for girls. It is not egalitarian. Boys have a broader selection of special skills they can hone, while girls learn virtues to nurture family harmony. Youths are selected by elders, often a grandparent, and sequestered in the woods for eight months to receive the most foundational and rarified teachings of their people. "People who have done traditional education will not sell the land," Yumte says. Fince Momo's second cousin, 23-year-old Urbanus Momo, confirms this and dreams that *wuon* will make a comeback. "The people who have done *wuon* know our heritage. They know exactly who you are, especially when it comes to land rights. They teach us to be wiser in solving problems."

Modern schooling and fundamentalist churches have taken their toll on Papuans' traditional education. Village schools provide a standard Indonesian curriculum and fail to make allowances for kids to learn in the forest for months on end. Fince Momo tells us that some families are adapting. When we spoke in 2020, a girl in Ayapokiar was sequestered and receiving traditional education from her grandmother while keeping up with school assignments dropped off by classmates. Almost all New Guinea Natives are Christian, with a pronounced sectarian difference in their attitude toward traditional learning. Yumte says that the Catholic Church has supported some efforts to revive *wuon*, while hard-line Protestants have labeled it devilry.

Something similar to Samdhana's work is happening on the other side of the island, on the north coast of PNG. A collective of clans in the Gildipasi region makes five-year pledges to protect their forests. These areas are the deep woods that the people don't use day to day, areas that in the 1980s and 1990s were heavily logged. "They logged right down to the riverbank," recalled Gildipasi leader Yat Paol when we talked by phone in 2020. "Some weeds and pests came, new species of toads and snails. These are strangers. They weren't here before. Also, now we have to go further inland to get materials for building." This is a problem, he said, for people who rely on forests for fish, game, nuts, canoes, medicine, fruits, building materials, and . . . Paol paused, puzzling over how to articulate a final item on this list. "I don't know what the word is, when we need to find peace, we go in there. We gain strength and we come back."

Three clans kicked out the loggers and in 2003 founded the protection pact. At first the agreements were written, reviewed by lawyers, and signed. Enforcement could involve beatings and court appearances. After ten years the clans began to shift to traditional forms of commitment. Now there is no written agreement. Instead, they celebrate elaborately on July 26 of every fifth year. Spiritual beings are greeted as parties to the agreements. "We believe that reality is not just physical. We believe that reality is both physical and supernatural," Paol explained. Spirits help with enforcement now. "It's serious business when we break contracts with the supernatural powers. . . . Anyone who breaches gets dealt with. We don't have rangers. The spiritual world is our ranger." He said that one death and two near-deaths have been attributed to spirits punishing noncompliance with the agreements in recent years.

Fear of spiritual consequences isn't the only motivation for adherence to the pacts. The protected areas spill resources into the surrounding landscape. "It's like a bank in there and things flow out," Paol said. "There's no Great Wall of China to keep things in! If a bird comes out, that's for us to eat." And, once every five years, the day before the contract renewal feast, villagers take plants, fish, and game from the preserved forests; that's one day out of 1,825 that reinforces the material and spiritual bond between people and place.

Paol recounted a similar story about the ocean, which in recent years has been incorporated in the pacts. The marine agreements are renewed two days after the forest commitments. He said that in 2018, "elders gathered in preparation and agreed that if a turtle was given by the sea between the 26th and 28th we will kill it ceremonially to seal the agreement. None had been killed in twenty years. On the 27th a turtle came up at 6 p.m. and that's not the normal time to lay eggs, so we knew right away this was sent by the supernatural world for ceremonial purposes."

The Gildipasi area is still small, in the thousands of acres, but membership has swelled from three clans in 2003 to over twenty in 2020. And the model is now being tried in a second, much larger region, the headwaters

A giant samaúma fruits in the Javari Valley, where it is treated as sacred.

of the Sepik, New Guinea's longest river and the most linguistically diverse place on Earth.

For our own good, the broader non-Indigenous society, which includes nineteen out of every twenty people on the planet, needs to live with a sense of respect and reciprocity with our ecosystems. Humanity's territory is Earth. We are aggregates of sky, incarnated courtesy of chlorophyll, water, earth, and other organisms, all of which we need to start treating like relatives. Joe Williams sees it happening. He introduces himself as a Māori must. First, he provides the names of his mountain, his river, his ocean, and his tribe; then, the names of eighteen human ancestors going back to the one who traveled 3,500 nautical miles by canoe in the fourteenth century from French Polynesia to the north island of modern New Zealand. Williams even knows the name of the boat: *Matatua*. Only after reciting this *whakapapa*, or genealogy, can he proceed to the lesser part of his identity, his job. Williams is a justice on New Zealand's Supreme Court.

Justice Williams, the first Māori on the Court, explains that in Polynesia, the sky is father, earth is mother, and ocean is a super-relative with exalted status among the 120 children of Earth and Sky. Seaweed, like people, have epic lineages worthy of songs. Williams sees traditional concepts starting to pervade broader society. In 2014, New Zealand's Te Urewera forest was legally transformed from property into a being with the same standing as an individual or corporation to seek relief from the judicial system. The Whanganui River has gained similar recognition as a legal "person" under the law. "The reason the politicians got it over the line in statutes here in New Zealand was that its time was right. It's not just the Māori community deeply troubled by the despoilation of the environment. It's the *entire* community," says Williams. "There's just beauty in this idea of us being finally recognized as the children of the planet."

7

Forests and the Real Economy

To save the planet-cooling megaforests, we need an economy that supports nature rather than erases it. That doesn't imply diminishing our economy; it means fully realizing it. As Partha Dasgupta puts it in his 606-page 2021 report to the British government on the economics of biodiversity, "To claim that there is an 'economic argument' on one hand and an 'environmental argument' on the other is to misunderstand the nature of economics."

The economy is what we call the sum of all our interactions with each other and the material world—including forests—to obtain goods and services in our quest for well-being. In a market economy, these interactions are fairly free. People buy and sell what they like from each other and use what they want from nature. Almost all communities enact rules to outlaw or discourage some kinds of market activity, such as prostitution, usury, and selling extremely addictive drugs or fraudulent or dangerous products. Most of us, however, still operate in economies that legally dice up forests until they're too small to get lost in. Our marketplaces permit commerce in forests' parts, with little concern for the integrity of the larger natural systems that make the parts possible. One of our central challenges in moving into a future with megaforests is to change that. If we fail, the results are predictable; large forests will be disassembled, just as they have been over the ages by the very economic dynamics that are at work today.

A little over 2,000 years ago, Julius Caesar marveled at the vastness of a forest he was conquering: "There is no man in Germany we know . . . who can say he has reached the edge of that forest . . . or who knows in what place it begins." This was the Hercynian Forest, whose extent Caesar himself and Pliny the Elder actually estimated to be a sixty-day march east to west and a nine-day walk south to north. The wooded area began at the Rhine River in the west and extended into modern Eastern Europe, encompassing all of today's Black Forest and then some. This was the canopy that remained intact in Caesar's time, after 4,000 years of agriculture and animal husbandry in the middle latitudes of Europe.

The Romans encountered forest peoples there whom they often admired for their attunement to the woods. Admiration notwithstanding, over the first several centuries of the Common Era, Rome beavered its way through the Hercynian woods and the smaller forests of what is now France, Belgium, Great Britain, Spain, and Cyprus. The trees were mostly burned, to smelt silver in Iberia, copper in Cyprus, and iron in France, Germany, and England. Wood warmed Roman baths and saunas, the popularity of which soared in the first century CE, and fired ceramic kilns, lime kilns, and glassblowers' furnaces. For a time, the Romans had plenty of beautiful dishes, vases, stucco, glazed windows, and relaxing soaks thanks to wood. It was too good to last.

The exhaustion of Spanish trees may have been the beginning of the end for Rome. Silver was smelted in Spain and minted into coins with profiles of one Roman big shot after another. Roman rulers were turning trees into money, but the trees ran out before the ore did. As emperors mixed progressively greater amounts of base metals into their "silver" coins, Rome's currency and authority both slid. By the early fourth century, many Romans had gone back to barter instead of using the wobbling monetary system.

After Rome, Central Europe's forests regrew to a modest extent during a period of migration, roving armies, and reduced rural populations.

Around 850 CE, however, the deforestation pendulum swung the other way and the medieval *grands défrichements*, or big clearings, reduced forest extent by a quarter or more. Forest cover in the temperate latitudes of Europe reached their current 40 to 45 percent by 1500. Today the continent has no intact forest landscapes outside of Russia and northern Scandinavia.

Around 1500 was also when Europeans first headed off to the Americas to start on two other megaforests, in southeastern Brazil and eastern North America. The Mata Atlântica, as Brazilians call their coastal rainforest, once flourished along the seacoast from present-day northern Argentina to the northeastern tip of Brazil. Separated from the Amazon by a thousand miles of savanna, desert, and swamp, the eastern band of forest was the size of four Californias.

When the Portuguese navigator Pedro Álvares Cabral first arrived in South America in 1500, his landing party's most immediately consequential discovery was a tree. The name for the tree, and ultimately the colony, came from the Portuguese word for "ember"—*brasa*—a reference to the wood's color. When boiled it produced a lovely red dye. *Pau* is a colloquial way to say tree, so the ember tree became the pau brasil. Thus was Brazil named after the wood that made Europeans' clothing red during the late Renaissance. Over time tenacious traders shipped out every pau brasil they could find, which was almost all of them. The species is now so rare as to have near-mythical status. But of course, the rout of the Mata Atlântica didn't stop there. Subsequent markets for sugar, coffee, cocoa, cattle, charcoal, minerals, and other woods erased the Atlantic forest in economic episodes spanning 500 years. Today, natural forest covers less than 8 percent of the Atlantic forest's precolonial 321-million-acre extent.

When the English came ashore in North America in the seventeenth century, they found towering oaks, beech, chestnuts, and cherry. The forest stretched from the sea to the Mississippi River and north to the Arctic. And the individual trees were giants, unlike anything on the settlers' home island. Tall, straight white pines were the most important find. In the 1650s, England's monarch, George I, was facing a mast crisis. Mast trees were long gone from his realm, and his merchant and military fleets relied on supplies from Baltic forests, which England's arch business rival, Holland, was threatening to cut off. With shipwrights struggling to make gale-worthy masts from multiple pieces of wood, the towering pines of Massachusetts, New Hampshire, and Maine quickly began to move across the Atlantic, along with lumber, tar, pitch, and oak for barrel staves. America was England's new woodlot. Native American forest economies, based on hunting, foraging, fishing, and small-scale agriculture, were forcibly supplanted by the English model of forest liquidation and larger farms.

The tree cutting continued apace in the century preceding American independence and with even less restraint thereafter. Sawmills, farms, and foundries obliterated what had been one of the planet's thickest woodlands by the mid-1800s; 70 percent of New York State's Adirondack forest fell between 1800 and 1885, despite the rough terrain. The nineteenth-century American iron industry, fired by charcoal, produced nearly twenty times as much metal annually as the English industry during its own charcoal-burning zenith between 1540 and 1750. In just fifty years, between the 1810s and late 1860s, Americans turned 200,000 square miles of forest into fuel. That's equivalent to all of the trees that once grew in Illinois, Michigan, Ohio, and Wisconsin. Metallurgy in America consumed 5 billion cords in this period, enough for seventy-nine stacks 4 feet wide and 8 feet long piled to the moon. The newly tattered American landscape was penned in by 3.2 million miles of wooden fences by mid-century.

In 1867, a scientist named Increase Lapham sent a report to the Wisconsin legislature stating, "From the facts already given above, it must be quite evident that clearing away the forests of Wisconsin will have a very decided effect on the climate." He warned of floods, erosion, and drought. Lapham even mentioned the carbon sequestration accomplished by trees: "They purify the air by absorbing carbonic acid gas." Lapham was talking about the local climate, not the global one. But he did ably point out the trade-off that Wisconsin (and North America more generally) was making, liquidating nature that was good for the community into salable food and the private paraphernalia of "civilized" life.

Looking at the eastern United States today, one might ask what the fuss is all about; there are lots of trees. Indeed, once Americans found richer farmland farther west and learned how to use fossil fuels, swaths of the eastern landscape resprouted. After 80 to 120 years, New England hardwood forests have been found to have carbon stocks on a par with old growth. The recovery of biomass, sometimes in lovely and useful forms such as the sugar maple, tells a story of hope. This is good. And yet, the precolonial forest did not "grow back." Today's forests are poor in pines and heavy in hardwoods. Beech abundance has dropped by around 70 percent. Their nuts and oaks' acorns were devoured in the billions by colonists' pigs. Beech trees, along with hemlock, were disadvantaged because they naturally grow back slowly, under the canopy of an existing forest, of which few were left. Chestnuts and elms have been erased by foreign blight. Wetland forests were filled for farming.

Most importantly, roads, towns, power lines, train tracks, mills, and mines grid the forests into slivers that can't sustain the plants, animals, and cultures that were present when English people first stepped onto the shores of Virginia. And some forests didn't come back in any form; since 1600, an estimated 310 million acres of forestland has been lost, an area 65 percent bigger than our current national forest system.

Given that intact forests give us rain, food, a healthy atmosphere, clean rivers, kaleidoscopic biodiversity, and feelings of primal spiritual connection, why do people behave as if we can't wait to get rid of them? Why do we turn the whole into parts? To avoid doing to today's megaforests what past generations did to theirs, it's useful to identify the economic dynamics that have reduced Earth's megaforests to the current five and, as we write, grind away at their edges.

Market theory describes how companies and people decide what to make and buy—how we interact with each other and the material world. The most fundamental plank of this theory is that when people exercise these decisions freely, society's overall well-being is maximized. Individuals and companies produce all sorts of goods that consumers want. You name it: corn, computers, yak butter, ivory, and violins, for example. Consumers send signals to producers by deciding what to spend their money on. Producers charge at least as much as it costs to make, grow, collect, carve, and bring their goods to market. The ideal marketplace has many sellers, so they compete and thus allow buyers to stretch their budgets and get higher-quality stuff. Producers whose costs are high, or who guess wrong about what people want, go out of business. This perfect market also has many buyers, so none of them has undue leverage over price.

In this model, with a given set of resources such as land, trees, workers, water, and minerals, the desires of people in a given society can be satisfied to the maximum extent possible. They vote with their purchases and signal to fellow humans how to use their time and other resources. No one is forced to do anything. It all kind of happens like magic, which is why, in 1776, the Scottish economist Adam Smith described the guiding power of the market as an "invisible hand."

Of course, the world we live in differs from this model of "perfect competition," as economists call it. Monopolies emerge and control prices. Consumers don't always purchase products that will make their lives better, spending on goods like OxyContin and huge sodas. Slick advertising can have as much to do with consumption choices as the inner voice of pure individual preferences. The model has ragged edges.

These are trifles when compared with the trouble it has with public goods.

"Public goods" is a technical term. The best place to understand it is out in nature, where public goods abound. In Brazil's Javari Valley Indigenous Territory, oxbow lakes are full of giant arapaima. The fish can reach the size of a dugout canoe and have roughly the same shape. Amazonian people will sometimes swamp their canoes so as to scoop a captured arapaima from the river. The Kanamary are standout fishermen and depend on the arapaima for a substantial fraction of their protein. A common sight in the Kanamary village of São Luis is two men shouldering a horizontal pole run through the gills of "small" arapaima whose tails drag on the ground as they traverse the settlement's soccer field.

When we visited São Luis, Korá Kanamary motored us around an oxbow lake just off the Javari River, pointing out hot fishing spots and

Arapaima caught in a lake near the Solimões River in Brazil.

(Photo by Marcos Amend)

lamenting the almost nightly illegal invasion of the lake by outside commercial gangs. Their clandestine catch supplies markets and restaurants as far away as the Peruvian city of Iquitos. Keeping the armed fishermen out, the Kanamary leader explained, would require the infeasible proposition of posting a night watch along miles of riverfront. Because access is prohibitively costly to restrict, the overexploited arapaima stock and its clean lake environment are at least partial public goods. In economic jargon, they are "non-excludable." Private goods, by contrast, can be kept exclusive with nothing more elaborate than a lock.

To qualify fully as public, goods must also be "non-rival." To experience a non-rival good, wait till it gets dark and then go outside and look at the moon. As intensely as you may gaze, the moon is undimmed for everyone else. Your use doesn't use any of it up. As with excludability, economists express this in terms of cost: once the good is created, the cost of providing it to one more person is zero. "Creation" of the moon view could be accomplished by regulating air quality or light pollution. Other non-rival examples include a fireworks display, public art, and—the mother of all public goods—a stable global climate.

Intact forests are overflowing gardens of public goods. The climate benefits, the wild, tangled webs of biological diversity, the pure waters that flow in their streams, and the variety of human culture are non-excludable and non-rival. Absent the heavy and very visible hand of government intervention, humans operating in markets aren't able to package and sell these things; few individuals will pay for clean air they can breathe for free. So, instead, we carry on packaging and selling things we can privatize from forests, like timber, minerals, game, and the land underneath the trees.

Megaforests are further undervalued, paradoxically, for the very trait that makes them so valuable to the stability of Earth's systems and thus to entire economies: their size. Standard economic thought values resources, and the products made from them, one unit at a time because things are more highly prized when they're scarce than when they are plentiful. This is called marginal valuation. It considers how much better off people are having one more unit of a good, or how much worse they are if a unit is subtracted. If there is a great abundance of something, the marginal unit is considered relatively worthless. When confronted with the stupefying abundance of trees, animals, and other resources in an intact forest, a blinkered but not atypical economist attributes low marginal value to each "unit."

In 1966, Ronald Reagan, then a Republican candidate for governor of California, questioned the need for a Redwood National Park that was proposed to protect the last of the species' old growth: "A tree is a tree, how many more do you need to look at?" When he said that, old-growth redwoods, which can live well over 1,500 years and mature into the tallest trees on Earth, occupied 1 out of 20 acres of their original range. The rest was, and remains, a forest of tree-children. Most redwoods today are cut as soon as they're several decades old in a forest that accumulates one-thirtieth as much carbon as ancient groves do and never replicates the old-growth ecology. Shaped by centuries of wind, fog, and fallen needles, old redwoods make a second forest hundreds of feet off the ground. Sitka spruce and huckleberry thrive in the canopy and host flightless animals, including worms and salamanders, that never touch the forest floor. These ancients eventually topple, often into creeks where they create deep pools for coho salmon.

From another perspective, that given voice by the gubernatorial candidate, the remaining giant redwood forests have an uncountable number of trunks, branches, needles, and roots. Viewed from the right hilltop, their canopy might stretch in a nubby green carpet right into the Pacific fog. This sense of abundance reflects the psychological difficulty humans have understanding large quantities. Most cultures count to a certain point and then jump to "lots." Who can tell the difference between a billion and a trillion? We use words like "countless," "endless," and even "infinite" to describe things that are finite and all too susceptible to exhaustion.

Meredith Trainor, head of the Southeast Alaska Conservation Council, says that such illusions endure in the United States' remaining megaforest state: "In the Lower 48 there's this understanding that resources are finite because they cut down so many of their trees and they saw the impacts to their salmon streams. Up here there's still a mix of naivete and willful ignorance that lets people imagine that just because you have this wash of forest outside your window, that we can just chip these little pieces off the edges, and it's not really a problem if we put in a road here and a road there because there's just so much of it."

Marginal value is often illustrated by the diamonds-and-water paradox. Diamonds are absolutely unnecessary and highly valued due to their scarcity. Survival is impossible without water, which is so abundant in some places that it's free. However, if a person is dying of thirst in the desert, that person is willing to trade a sackful of diamonds to get a drink. The seeming abundance of wood, creatures, and carbon in megaforests is like that of water in a wet place. It appears sensible to—in Meredith Trainor's words—"chip these little pieces off the edges" of a big forest to grow something or mill boards or make way for a house. Where nature abounds, the cost of using up any small bit of it appears trivial.

The logic of marginal analysis is the very core of traditional economics, and it does roughly describe human behavior in a free market. If you're a coffee drinker, think about how much you value your first cup in the morning, versus the second, third, and so on, until the cup someone would have to pay you to drink. But economics is not just a system of observation. It is also used to prescribe, to make policy, to decide if a forest should be protected and, if so, how much of it. Thinking marginally, the market—and the leaders who choose not to restrain it—pare and dice big forests until their contents are scarce. In the most extreme cases, marginal values are treated as low until populations are down to numbers that would fit on an ark.

The economy, again, is how humans use resources to survive and make ourselves happy. An economy that incinerates public goods and overlooks the value of large-scale ecosystem integrity is failing. Destruction to date has sometimes been an inadvertent by-product of economic activity and sometimes the intentional project of pioneers, emperors, and entrepreneurs. Until recently, in both scenarios, most people have been unaware that the production of stuff has been eroding the biophysical foundations of our economic system that produces stuff. Forest fragmenters of the past were innocent of an awareness we now possess: this can't go on. Our planetary and economic health is incompatible with more forest liquidation. For starters, keeping carbon in, and on, the ground is undervalued in the extreme.

To make a place for Earth's great remaining forests in the kind of economic framework that orients governments and development banks, we must reconceive what we mean by the margin. Because there is a very small number of very big places whose contributions to climate, biodiversity, and culture can't be replicated any other way—because, in other words, plenty has become scarce—the unit of marginal analysis must be the forest, not the trees. Rather than thinking of trade-offs involving the next acre, or even 100,000 acres, our economics should consider the entire forest. Better yet, the entire megaforest, with its core of IFLs and encasing buffer of ecologically functioning, if somewhat more intensively used, woods.

Forest-oriented metrics should be brought into our mainstream discussion of how well the economy is doing its job—allocating resources to achieve human well-being. Here's what our policymakers and economic reporters discuss now: the monetary value of everything we produce, also known as gross domestic product (GDP); the annual growth rate of GDP; the present value of expected future profits of a collection of publicly traded companies (stock indexes like the S&P 500); the share of willing workers who have jobs (the employment rate); whether our society is selling more stuff to others or buying more (balance of trade); and the psychological readiness of people to buy (consumer confidence). In other words: a fury of transactions, a pile of profits, human toil, a trade edge over other countries, and an inclination to buy. Those are our main proxies for well-being. The *New York Times*, in a bid to expand the definition of the economy, presented ten different indicators in a retrospective on the economically tumultuous year 2020. Not one was environmental. All of them could be improving in a world that is eating itself for breakfast.

During the coronavirus pandemic, people flocked to nature in record numbers to lift their spirits. But the forest and other ecosystem losses that erode our daily well-being and even threaten the future of the stock market and the GDP are treated as extracurricular subjects. Economic expansion in any legal industry is reported as unequivocally good. So, we invite reporters to ask, in every story they cover, whether the news is good for nature. Media outlets might include climate costs and benefits in all business reporting. Data are available to gauge the atmospheric consequences of construction, oil drilling, consumer spending, air travel, and shipping. Why not show it? Another positive step would be a quarterly forest report on losses and gains, fragmentation, carbon uptake and emissions, and changes in protected status. Global Forest Watch already has most of the information online; it should be covered as breathlessly as any conventional measurement of the economy. Transparency will make consumers more thoughtful and encourage government and business to score better on forest indicators.

Economic analysis can inform but shouldn't overwhelm common sense on forests and climate policy. For instance, many early economic analyses of climate change concluded that decision-makers should take care not to overinvest in stopping global warming, prescribing a cautious ramping up of action. That's because cost-benefit analysis discounts the value of future benefits, simulating the fact that people tend to care more about the present than the future. We generally require an interest payment to put our money in the bank and hold off on using it until later. The interest rate is used to do the math of discounting and has a huge impact on what economists conclude should be done about climate change. A moderate-to-high interest rate means that society places much more value on benefits/consumption now than in the future. If one doesn't care much about the future, there's not much economic rationale to do things now, like saving forests, to secure it.

The most famous cost-benefit analysis of global climate action was done in 2006 by Sir Nicholas Stern at the request of the British government. Stern recognized the timing dilemma; conventional interest rates would drive the value of saving the climate so low that little action would be economically justified, despite a real risk of catastrophe. He used a rate of 1.4 percent per year, a half or even a third of rates economists typically apply. Stern's choice was grounded in the notion that social policy shouldn't value today's people more than tomorrow's.

His study concluded that it was worthwhile to invest heavily now to avoid ruined crops, deluges, famine, and the other future consequences that will come from climate inaction. Some economists felt he cooked the books with his low rate and have come to more ambiguous conclusions about spending to avoid climate disaster. Stern, whose 575-page report is still the only global study of its kind, was unrepentant ten years after its release. He said, in fact, that he underestimated the benefits of action.

Stern came to the correct conclusion by retrofitting cost-benefit analysis to do something it wasn't built for. He got the world's attention, which is good, but gave a mistaken impression that these economic calculations, done correctly, could guide the long-term management of the planet. In fact, cost-benefit analysis was brought into policymaking for much more limited purposes. It debuted in the 1960s to evaluate (and rein in) US spending on weaponry. It spread to environmental policy, primarily to decide on human health and safety regulations over time scales of a few decades at most.

Two-century analyses like Stern's are unavoidably rife with guesses about what will happen and what people will care about a long time from now. And they can never price all of the environment's intangible, aesthetic, and spiritual values. Above all, though, the problem with cost-benefit analysis is that conservation, like the planet, is supposed to be forever, not just for the next 20 years or even the next 200. Unfortunately, in cost-benefit analysis, forever is worthless. Present value is a fraction with a fixed numerator and a denominator that balloons over time. No matter what interest rate is used (as long as it's positive), present value's trajectory is eventually an asymptote to zero. Any notion of how humans should live in this world over the long run needs to be moored to a sturdier rock than an equation that, within a thousandth of human history, makes the world disappear.

The best uses of economics in environmental policy are humbler, focused on how to act most effectively in the short and medium term, selecting from a range of policies, technologies, and geographic priorities to accomplish a goal that's been fashioned from various streams of knowledge and ethics. In the case of climate change, the United Nations Framework Convention on Climate Change (UNFCCC) has already set a goal—staying below 1.5°C of warming. Economists can help by informing decision-makers on the costs, benefits, and equity impacts of their various options, including megaforests, for getting there.

Enlightened economics will help countries fashion forest policies that are realistic about markets' limitations, that avoid irreversible natural and cultural losses, and that are geographically specific and expansive.

When we refer to forest policies, we are talking about agreements within a community to put some fetters on how supply and demand operate in nature. Rules are as old as humanity, and rules on using nature are among our oldest. Policies are how groups put the market within a box of ethics and shared purpose. Agreements can be struck with a handshake or eating a turtle ceremonially in a tiny Papuan village. They can be codified in thousand-page national laws or set forth in treaties embraced by multiple nations.

The Paumarí people of the Amazon, for example, explicitly agreed among themselves to suspend fishing in their overharvested lakes. This policy restored the stocks and produced monster arapaima that became a regular part of Paumarí livelihoods. Mosetén villages in Bolivia made a similar arrangement, limiting hunting to quantities for personal consumption, which kept the forests flush with native deer and wild pigs. In a labyrinthine Amazon-biome wetland of multicolored waters known as the Fluvial Star of Inírida, Colombian communities who capture ornamental fish for the aquarium trade devised rules to limit catches and ensure both ecosystem health and their long-term incomes. There are countless stories like these among small groups of people who have a common set of values, who police each other with social pressure, and who know very quickly when the contract has been broken. Describing these dynamics won Elinor Ostrom the 2009 Nobel Prize in economics.

But there are also plenty of examples of provinces and nations that use environmental laws and nature reserves as collective responses to "produce," in an economic sense, clean air and water, game and more obscure wildlife, old trees, and the continued existence of forest-linked cultures. The three biggest megaforest countries, Brazil, Russia, and Canada, have already done remarkable work to protect the woods. The United States has cleaned up its air and rescued species such as wolves and bald eagles. In 1980, the US federal government carefully selected 157 million acres of Alaska—37 percent of the state—to be conserved forever. And Congo basin countries have made strides to save their great apes and forest elephants. These are all public goods saved from the devouring maw of markets by the cooperation of millions of humans.

Internationally, we have market-limiting agreements such as the Great Lakes Water Quality Agreement, the Convention on Migratory Species, the Convention on Biological Diversity, the United Nations Declaration on the Rights of Indigenous Peoples, the UN Convention to Combat Desertification, and the UN Framework Convention on Climate Change. Best results come when global accords cascade through ever more local agreements, all the way to deals upheld by people in the forest.

Policy strategies to save the megaforests can include across-the-board incentives for conservation. Among the best known of these economic "nudges" have been certifying sustainably harvested timber, restricting trade of commodities produced on deforested land, granting preferential access to agricultural loans, and rewarding forest stewards with lower taxes or cash subsidies. These incentive-based approaches telegraph a social goal into the marketplace rather unobtrusively and appeal to landowners, who are encouraged rather than commanded to conserve.

But it's not enough to nudge the economy in the right direction and hope for the best in terms of forest landscapes. Societies must explicitly protect woods that are big and healthy. Forests' climate-saving power depends on the intactness of the whole forest system. A patch surrounded by pastures will hold much less carbon, and with far less security, than the same area embedded in a larger natural ecosystem. Size and location matter for water, fish, game, sacred sites, local foods, ancient trees, culture, and scenic beauty. Similarly, there are poorly understood tipping points, such as the precise degree of deforestation that could turn much of the Amazon into a land of grass and shrub; an incentive may or may not encourage enough forest protection to avert catastrophe. And some damage to ecosystems is irreversible—you can't cut a forest with endangered species or uncontacted people one day and then change your mind the next morning when you see the trouble you've wrought.

Megaforest countries already have made a good start of protecting specific places. They all zone forests in some fashion for uses such as parks, water supplies, timber, roads, and settlements. That's a solid base on which nations can build, making the spaces for forests bigger and more secure from roads and other reversals. The nations who do this work are giving to the global commons by producing climatic and other public goods for us all. Which raises a question: should they, perhaps, be getting paid?

8

Money Trees

In 1987, two tropical biologists and a Yale economist walked into a jungle. Charles Peters, Alwyn Gentry, and Robert Mendelsohn (the economist) wanted to know whether harvesting fruits, nuts, and fibers from an Amazonian rainforest was profitable. They concluded it was. Sustainably harvested forest fruits and rubber generated six times as much income as logging for the small community of Mishana in the Peruvian rainforest. Publication of the findings in the journal *Nature* in 1989 kindled a hope that forests would protect themselves. Many conservationists spent the next decade seeking ways to help countries and communities cash in on ecotourism, Brazil nuts, rubber, rattan, and even careful timber extraction.

But there was one line in the Mishana study that many overlooked: "not every hectare of tropical forest will have the same market value as our plot in Mishana." That was a profound understatement. The plot in Mishana was fruit rich and a mere 18 miles from Iquitos, a big city whose citizens are partial to perishable jungle produce. Conditions were ideal for the wild fruit business. Sustainable forest production has provided strong incentives for conservation in places like Mishana, small areas endowed with access to both product and consumers. These successes have given us something to celebrate. But the same economic approach hasn't worked to protect really big forests, which, by dint of their bigness, always have massive parts far from markets.

Then came carbon. In contrast to the delicate fruits of Mishana, carbon delivers value without being moved. If carbon storage could be monetized, conservationists began to realize, it would provide a financial infusion to governments and communities that protect intact forests. This was, and still is, a tantalizing possibility.

Carbon finance has focused primarily on tropical rather than boreal forests for three reasons. Tropical rainforests quickly accumulate massive amounts of carbon above ground level, packed in stems, bark, leaves, ferns, creepers, flowers, and other plant matter. While boreal forests do contain the world's richest carbon caches, up to 90 percent of it is underground, where it may or may not be released by activities on the surface. The risk of a boreal greenhouse gas hiccup is hair-raising, given how much carbon is down there, but the vulnerability of subsurface northern carbon is not yet as easily quantifiable as is conversion of aboveground tropical plant parts into CO_2 gas. Further, most of today's outright forest loss and degradation is happening in the tropics. Finally, the tropical countries with large forests are low- and middle-income nations with extremely tight budgets; they are working to extend basic services like health and education to their citizens. Most of these nations have contributed very little to the carbon accumulated in the atmosphere so far. Brazil and Indonesia are the only two on the global top-ten list of historical emitters. The tropical forest countries need and generally deserve help with the costs of keeping carbon in the biosphere.

The idea of paying forest-rich countries for carbon stored in their wooded zones first manifested in a deal in 1996, when The Nature Conservancy, British Petroleum, and American Electric Power paid the Bolivian government $9.5 million to cancel logging on 1.5 million acres of rainforest and add the area to the Noel Kempff Mercado National Park. The parties calculated that a million tons of CO_2 emissions would be avoided over a ten-year span. This arrangement was entirely voluntary. Neither British Petroleum nor American Electric Power was required to reduce emissions of greenhouse gases. They did it, at least in part, to showcase corporate responsibility. The Nature Conservancy participated to protect a large intact forest of surpassing interest for its biodiversity. But all three buyers and the seller did carefully quantify and agree on ownership of credits (split 50-50 with the Bolivian government) for the emissions reductions, in case such things turned out to have monetary value later on.

That was one year before the first binding climate treaty, the 1997 Kyoto Protocol to the UN Framework Convention on Climate Change. This agreement categorized the world into rich countries and developing countries, mandating limits on emissions from the rich ones. It also set up the Clean Development Mechanism (CDM), which allowed wealthy countries to do exactly what The Nature Conservancy, British Petroleum, and American Electric Power had done in Bolivia: pay for emissions reductions in less-rich countries. But the CDM didn't help out those three organizations as far as their speculatively created carbon credits were concerned. Largely focused on industrial emissions, the CDM only recognized forest credits from newly grown or replanted forests. Saving a standing forest, despite being cheaper and more environmentally beneficial than planting, didn't count. In fact, building a new hydroelectric dam that flooded ancient forest could earn CDM credits if the project was replacing a fossil fuel plant. Many CDM applications of this sort appeared in the early 2000s.

The logic of saving forest to save the planet, however, is so compelling that at least 350 voluntary projects in the mold of the Noel Kempff Mercado arrangement sprang up over the next twenty years. These deals, though generally small, were proving grounds for how to structure agreements, measure their results, and make sure they didn't simply push forest loss somewhere else. As economists would expect, spending extra money voluntarily to protect forest carbon, a public good, did not take the corporate world by storm. Some companies volunteered while most watched, and 90 percent of funds to save forest carbon continued to come from government. In human communities on the scale of nations, or the planet, volunteerism can only be part of the solution.

Paulo Moutinho understood that saving rainforests had to become part of the official UN climate approach for carbon finance to have a discernible impact on the Amazon and other tropical jungles. Cofounder of the Amazon Environmental Research Institute (IPAM), Moutinho teamed up with Márcio Santilli, from a Brazilian advocacy group called the Socio-Environmental Institute, and Steve Schwartzman, an anthropologist with the US-based Environmental Defense Fund. They started attending the annual Conferences of the Parties (COP). These are UN meetings during which member countries convene to hash out details of the climate treaty, and thousands of organizations put on a parallel climate expo and lobbying festival. Moutinho and his allies promoted an idea they called "compensated reductions" of deforestation—getting paid to deforest less. The proposal was bitterly resisted by the Brazilian government, which didn't want the spotlight on Amazon destruction, as well as by environmental groups like Greenpeace and World Wildlife Fund. They contended that deals like the one in Bolivia would allow rich countries to continue polluting with no guarantee that the forest "offset" would reduce overall deforestation.

"We would get kicked out of meeting rooms and literally see people dumping stacks of our publications in the trash," Moutinho reminisced with a smile. "We had to have our side events in hallways and the dark corners of bars."

Then, in 2003, Marina Silva, from a family of rubber tappers whose living depended on an intact Amazon, became Brazil's minister of environment. That year at the climate COP, Moutinho's forest carbon group didn't get kicked out of any meeting rooms. Their event was attended by 600 people, including officials from Brazil's foreign ministry. With a charismatic cabinet minister in their corner—plus fast-improving technology to measure the carbon in forests—they had the wind at their backs, and in 2007, the governments of Costa Rica and Papua New Guinea successfully argued that forest conservation in developing countries is a legitimate climate-saving action and that rich countries that help pay for it should get credit. Thus was born an acronym that has dominated over a decade of forest finance discussions: REDD+, Reducing Emissions from Deforestation and forest Degradation. The "+," appended later, kept the acronym short while embracing various activities, such as careful forestry, that could secure and add carbon to forests.

A potentially massive economic incentive for tropical forest conservation hove into view. On the horizon was yet another UN climate conference. In Copenhagen, in 2009, many hoped and expected that nations would agree on new binding limits on emissions, including a global limit that would avoid apocalyptic warming—important not just for planetary survival but to uncork funds for intact forests.

Such a limit would serve as the "cap" in a cap-and-trade system, a policy proven to work, for example, on sulfur dioxide emissions in the United States. Sulfur dioxide (SO_2) comes mostly from burning coal and causes acid rain, which is bad for rivers and forests and dissolves anything, such as statues and buildings, made out of marble or limestone. In 1990, new rules under the Clean Air Act set a cap of approximately 9 million short tons of sulfur dioxide emissions per year, a reduction of more than 50 percent. Existing power plants received shares of the 9 million tons of allowable emissions. But some companies could reduce their emissions more cheaply than others. For example, some had access to low-sulfur coal from Wyoming, while others depended on high-sulfur varieties from Appalachia and had to install expensive scrubbers to catch the pollution as it left the smokestacks. Power plants that could reduce their emissions cheaply sold allowances to companies whose abatement costs were higher. Allowance trades happen in a cap-and-trade system because abatement costs vary and because allowances are rendered finite by the cap. So dramatic were these efficiencies in the case of sulfur dioxide that cleanup costs were 18 to 27 percent of the figures estimated before the program started.

Protecting carbon in forests is cheap relative to the costs of reducing emissions from energy, cars, and industry. In a global carbon cap-and-trade agreement, therefore, forested countries expected to be sellers.

The Copenhagen meeting was a bust. Amid global financial meltdown, grand gestures were not forthcoming. China sent second-tier functionaries to final negotiating sessions and resisted specific strong targets, even by other countries. Not even a highly motivated President Barack Obama, negotiating in person, could save it. The countries agreed on almost nothing. Neither a global cap nor country-level responsibilities were established, so an international market in carbon reductions was off the table. Money for saving carbon-rich forests would have to come from somewhere else.

Somewhere else has turned out to be payments from governments that decided to press on with the twin quest of climate and forest protection despite the fact that they would not be getting credits in return for their money. These deals had begun to emerge as pump-priming experiments three years before the Denmark conference. After the Copenhagen debacle, they continued as the only game in town. Norway led the way. By 2014 it had pledged $4 billion, four times as much as the United States and more than nine runner-up donors combined. Norway and others made early deals with Brazil, Indonesia, Guyana, Colombia, Ecuador, Peru, Liberia, and Tanzania. The arrangements have paid forest-rich countries to prepare and, in a few cases, enact programs to reduce deforestation and forest degradation.

These experimental deals have, as one might expect, turned up areas that require adaptation. Some argue that the deals could be more straightforward. Donor countries that worry about funds being misused have stipulated how the forest-rich countries should spend the money. Mirey Atallah, a Lebanese climate expert and the chief technical advisor for the DRC's REDD+ fund, FONAREDD, thinks this is a mistake. She believes that REDD+ would work a lot better if the countries paying for forest carbon left the details to the countries where the trees are. "This is a market transaction that involves a service being delivered and a payment for that service. Donors are still conditioning the payment for that service to how the money will be spent. You know it's as if you were buying a phone from Apple and you were asking Steve Jobs or whoever his successor is, 'How are you going to spend that money?' You're paying the money to purchase the phone because the phone satisfies your requirements. There's no reason for you to go any further than that. Whether he buys a bottle of whiskey or spends the money on his child's education is none of your business. However, what we see in the context of REDD is there is still conditionality and a requirement to know how the country is going to use the funds."

One evening in the town of Sausapor, West Papua, Yunus Yumte discussed REDD+ as we waited for dinner in a two-table restaurant the size of a bus stop. He was involved in the Papuan REDD+ strategy, which slots into the larger Indonesian plan for cutting forest emissions. "We've been designing our REDD+ system since 2011 and the emissions are just going up." The countries helping fund Indonesian REDD+ are trying to do the right thing, Yumte said. But too many years have gone into deciding how much carbon to preserve in which provinces and how to measure the gains with precision. He pantomimed awarding contracts to expert consultants. "Here's a million for you to design the MRV system," he said as he handed his colleague Sandika Ariansyah a pretend sack of money to come up with a bespoke measurement, reporting, and verification protocol. "And here is two million for your organization for . . ." he paused, with a finger on his chin, "a baseline study!" he said, handing imaginary cash over to John.

Because of the way REDD+ funding has been set up, much of the money has gone through aid agencies with zealously careful procurement policies and little technical experience with forests or climate. The financial peristalsis was so gradual at the World Bank and the UN that of $2.2 billion pledged for REDD+ finance by 2006, only $247 million had been disbursed eight years later. The two were hardly alone—country-to-country money also moved slowly. Further, Mirey Atallah points out that while the government of the DRC is on the hook for hitting forest protection benchmarks, the money is actually spent by the development agencies on a jumble of projects that are often not coordinated with one another or with the host government.

Another critique of REDD+ funding is how little of it there has been. Norway is the world's twenty-sixth largest economy. And yet, on a bar graph showing countries' contributions of forest carbon funding, Norway's bar looks like it wandered into the wrong chart, towering over the incongruously flat rectangles representing the other countries' giving. Because contributors have been few and their giving timid, tropical forests received just 3 percent of total climate funding during 2010–2015 worldwide, despite potentially representing 30 percent of the near-term solution.

Lee White's experience at the center of forest policy in Gabon has left him doubtful about REDD+ funding. He's minister for Forests, Sea, the Environment, and Climate Plan. A biologist with decades of field experience, White has been charged by gorillas and elephants multiple times. When we talked in early 2020, he had recently witnessed a near-death encounter. His wife was hoisted and trunk-hurled by a forest elephant in one of Gabon's national parks. But he looked at home in white shirt and black tie in front of gold-fringed drapes in his Libreville office.

"In the last fifteen years we've reduced our carbon from the forest by about 350 million tons. Gabon has done about $2 billion of REDD at $5 per ton," said White with a wry smile. $1.75 billion, to be precise, in free planet cooling from a country with a GDP per person less than a fifth of that in the United States. Had they been paid for it, "Gabon alone would have sucked up almost all the REDD funding that has been available for the last ten years. And that's Gabon. We're just 10 percent of the African rainforest and the African rainforest is only half as big as the Amazon rainforest so I'm quite a REDD skeptic, even though I go to those negotiations and have done for ten years."

The $5 figure White cites is the value that has been assigned to a ton of carbon saved by tropical forests in almost all the national agreements struck to date, most of them with Norway. "I don't even know where the $5 came from, to be honest," says Mads Halfdan Lie, the policy director at Norway's Ministry of Climate and the Environment. A couple of academic models in 2008–2009 did find that most deforesters would theoretically be dissuaded with an incentive between $5 and $8, but when stored carbon debuted as a real-world commodity, it had no track record, so no one really knew what it was worth.

Fifty dollars was the average answer of 365 economists surveyed in 2015, based on the cost of the damage that a ton of carbon does in terms of floods, draught, fire, and pestilence. This is known in policy jargon as the "social cost" of carbon. It's a theoretical number that would be a useful benchmark for carbon pricing if officials, not markets, were to set the financial value.

A competitive market, in contrast, should settle on an equilibrium price somewhere between the forest owner's cost of protecting a ton of carbon, on the low end (what the 2008–2009 studies estimated), and the buyer's cost of preventing their own emissions—for example, by scaling back polluting activities or switching technologies—on the high end. To be a "competitive" market, however, buyer and seller would have to have a similar amount of leverage in the transaction. Norway has, to this point, been virtually the only forest carbon buyer, while many tropical countries are selling.

Though Norway does not appear to be abusing that leverage—it has been pursuing deals that are fair and effective—if fifteen to twenty countries were buying, a more reasonable market price might prevail, and we would start to see the sort of expansive forest protection that's needed. There are, for example, fourteen countries that pump more oil than Norway. All are absent in the carbon market. There are twenty-six countries with higher per capita CO_2 emissions, including the United States, whose pollution per person is approximately double Norway's. High-emitting, oil-rich, and money-rich Middle Eastern countries are also missing in action, as is China.

Lee White's take on the actual going rate of $5 per ton of avoided carbon emissions? "We don't believe that the REDD+ mechanism at $5 a ton is going to encourage anybody to save their forest. By making our forests economically important and contributing to the economy of Gabon, that's how we think we can make the forests survive. REDD might help us manage our national parks, but it won't help us develop our nation and give jobs to our young people."

Further, White pointed out that when trees are kept standing to save the climate, they also host chimps and elephants and circulate rain around continents. "That a ton of pristine rainforest carbon is only worth $5 per ton is something I take offense at. I think rainforest carbon should be worth a lot more than reducing carbon from a cement factory in Europe." What's the right price? "It should certainly be above $50."

Mirey Atallah agrees that $5 is laughable. She pegs the true costs of reducing a ton of rainforest emissions in the DRC at $30 or more.

Notwithstanding White's REDD+ criticisms, Gabon and Norway made a REDD+ deal in 2019. Norway will pay $150 million over ten years for Gabon to reduce forest emissions by 15 million metric tons. The deal represents several big improvements. Gabon will be paid for carbon that its jungles absorb during that time period, not just for avoided deforestation emissions. This is important because Gabon is a country with almost no deforestation to avoid. It was one of eleven countries identified by a 2007 paper (entitled "No Forest Left Behind") as "high-forest, low-deforestation," or HFLD. In the past, REDD+ buyers wouldn't pay to avoid deforestation where it wasn't happening. In a way, this is an eminently sensible position and crucial where saving forests was envisioned as an offset for buyers' own emissions.

The downside, however, has been that remote forests and whole countries with low deforestation have been sidelined in carbon funding. They include some of the poorest countries, which need funds to build defenses against the eventual arrival of pressure on their jungles. The absurd result was that a country such as Gabon, Suriname, or the Republic of the Congo, with historically low forest loss, would need to start deforesting in order to establish the threat. Then they'd get paid to stop it. That scenario raised the question of whether the period of high deforestation could be locked in forevermore as an advantageous baseline on which the country would have little trouble improving—or whether a moving historical baseline, which some countries have used, would be applied, requiring it to seesaw between deforestation and preservation to create salable reductions. The HFLD countries have tended to make the compelling argument that they should be rewarded for having lasted this long as forested nations, not compelled to develop a deforestation habit to start attracting REDD+ payments. Gabon's transaction doesn't entirely reach that ideal, but it's a step in the right direction.

The Gabon-Norway deal also increased the price to $10 and defined it as a floor, not a final figure. Lie explains that Norway guarantees Gabon $10 per ton and holds the credits. If Gabon finds a buyer willing to pay

Congo forests have plant carbon and chimp babies.

more, then Norway will release the credits to the new buyer. For example, if Gabon gets an offer of $25 per ton from Spain at some future date, Norway will transfer the credits to Spain and Gabon will pocket the additional $15 per ton. The carbon benefits will be made more attractive to buyers by following a rigorous new standard called The REDD+ Environmental Excellency Standard—or TREES. Lie says the idea is to encourage "super-high-quality" emissions reductions in advance of whatever carbon market may eventually emerge.

One of the most impressive things about Lie and his Norwegian colleagues pursuing these REDD+ deals is the equanimity with which they absorbed the arrival of Brazilian president Jair Messias Bolsonaro, who rolled onto the international scene like a stun grenade of anti-environmental bombast. When Bolsonaro took power on January 1, 2019, he declared an all-out offensive on the Amazon. While Brazil's once-stellar record of megaforest protection had already begun to reverse under Bolsonaro's predecessors, Michel Temer and Dilma Rousseff, the new president made Temer and Rousseff look like John Muir and Rachel Carson. He set out to loosen environmental licenses, shrink protected areas, cut new roads into the forest, and allow agribusiness to rent Indigenous peoples' forests and turn them into soybean fields. Deforestation spiked.

His environment minister, Ricardo Salles, had never visited the Amazon before assuming his post. Salles proposed to redirect the $1 billion Amazon Fund, supplied in a REDD+ deal by Norway and Germany, to address urban environmental issues unrelated to climate change, while federal authorities made war on the rainforest. It was a naked provocation. The European donors resisted, and the fund was frozen. One could conclude from this jousting that the scheme worked exactly as designed; Brazil stopped performing, and Norway and Germany stopped paying.

That's how Lie sees it. He is an ultra-runner who likes to lope as far as he can out into the Norwegian woods, sling a hammock between two trees, spend the night, and run back in the morning. Endurance and patience are useful virtues in his line of work, too. He says his compatriots still support REDD+ despite Bolsonaro, despite the nonemergence of an international market, and notwithstanding corrupt governments and bureaucratic intermediaries with the agility of cold molasses. "We warned all the politicians ten years ago," he said when we spoke, at the end of the new Brazilian president's first year in office. As a result, setbacks have been greeted with frustration but not shock. Remarkably, and to their immense credit, Norwegian political leaders across the ideological spectrum continue to favor paying for tropical forest. "People just seem to like forests. It's something that's close to people's hearts."

The fracas with Bolsonaro uncovered a risk that future deals would do well to address. If a forest carbon–selling government has no reasons of its own, over and above the payment, to keep the forest, any deal will be vulnerable.

The Amazon Fund money supported environmental and Indigenous organizations and public agencies whose purpose was to save the rainforest and in fact accounted for 75 percent of all overseas support for Brazilian science, advocacy, and public agency work to conserve the Amazon in the 2010s. By pouring sugar in the gas tank of the Amazon Fund, Bolsonaro cut off money to his environmental critics at no great cost to Brazil's economy. Brazil's 2019 federal tax receipts were approximately 3,840 times the $100 million Amazon Fund income lost by its environmental agencies, nonprofits, and universities that year. The fund could easily get lost in Brazil's fiscal sock drawer and not be missed for a long time.

Public opinion and the mainstream media in Brazil are sophisticated and favorably inclined toward protecting the Amazon. It is the top country in the world in terms of public recognition of the term *biological diversity*. A 2021 UN-Oxford public opinion poll on climate change asked which of eighteen listed solutions people preferred to address the crisis. Forest protection was the top choice among Brazilians, favored by 60 percent of respondents. The derailment of the Amazon Fund despite this majority support reveals that certain elements of government and business, especially those who have power over the fate of the forests, need to care more about the success of the carbon deals. For instance, ministries of agriculture, planning, transportation, and energy have that sort of power and influence but little role in REDD+ initiatives. They don't get money from the Amazon Fund. When political winds shift, a national initiative to preserve forest carbon would have greater chances of survival if it engaged the agencies that otherwise may be the most eager to see it disappear.

Despite the setbacks, carbon-saving agreements between countries rich in forests and countries rich in money are essential. Paulo Moutinho gave some compelling reasons to keep trying. "The Brazilian government learned that you can set targets for forest protection, just like targets for health and education. And these forest targets became part of our national climate law." The Amazon Fund has fueled a creative explosion of science and low-carbon development projects. It has provided a blueprint, he said, for the $10 billion global Green Climate Fund and for a German REDD+ initiative with state governments. It enabled the real-time Deforestation Alert System launched by the Brazilian group Imazon, and funded the remarkable MapBiomas site, which tracks changes on every acre of Brazil from 1985 to the present. Moutinho called MapBiomas "probably the biggest independent land-use monitoring system in the world." On the ground, he added, intensified agriculture projects enabled some small farmers to increase profits by 120 percent while simultaneously cutting their forest emissions by 60 percent.

"Now we have the recipe," Moutinho said, citing Brazil's successful package of law enforcement, environmentally conditioned agricultural loans, and new protected areas that, together, smothered deforestation.

The Amazon Fund was not simply a quid pro quo of cash for environmental performance. It was a modest financial boost and international show of support for homegrown Brazilian anti-deforestation efforts undertaken by government and activists that kept billions of tons of carbon out of the atmosphere in the early 2000s—at a time when beef and soybean production boomed. And the country made irreversible gains in the science and talent needed to protect the forest.

Carbon is not equal to the value of the forest. But its safe storage is crucial to humanity's shared destiny, and carbon is produced and stored in intact forests in terrific quantities. It is, furthermore, unsurpassed as a conveniently uniform commodity all forests offer—from Mishana to the middle of nowhere. Funding indexed to carbon can help pay for a variety of interventions that will help save the forest: guardians, protected areas,

Chestnut-eared aracari perches in a carbon-dense forest near the Solimões River, Brazil.

alternative non-road transportation, subsidies for sustainable products, and transition-softening programs for people now employed in high-emitting lines of work. As more buyers step into this arena, spending could climb into the tens of billions of dollars annually in agreements that emulate and learn from "buyers" like Norway and "sellers" like Brazil and Gabon, the pioneers in financial cooperation to save big forests.

9

The People's Forest

Only one Russian social movement survived the 1917 revolution, Stalin's purges, the Cold War, the collapse of the Communist economy, and a 2015 law that allows government to prosecute nongovernment organizations vaguely defined as "undesirable." That movement is nature conservation. In a sense, it started with Tsar Nicholas II. The last Romanov monarch took power reluctantly at age 26 after the untimely death of his father and ruled for twenty-two tumultuous years, often with the mystic Rasputin at his elbow. The tsar was forced from power in 1917 by the Bolsheviks, who executed him with his family in a basement the following year. Amid all this well-known drama, a little-known official act—one of ill-starred Nicholas's last—secured him a legacy that may be his most important. He created Russia's first nature reserve. In doing so, he launched a protected area system based on science, a global first.

The Barguzinsky Zapovednik covers 613,000 acres of forested mountains on the eastern shore of Lake Baikal. Nicholas's primary motivation for creating it was a small carnivorous tree-climbing member of the weasel family called the sable. Sables have the most prized fur of any animal in Siberia, the invasion of which was driven, in no small measure, by the demand for fur in Europe. People had been trapping and shooting the cute brown creatures out of the trees for centuries, and the sables were almost gone by the 1890s, when then-prince Nicholas, an avid hunter, toured the great eastern forest.

The idea of protecting nature had been percolating for a couple of decades in Russia. Its most prominent proponent was a bee expert at Moscow State University, Grigorii Kozhevnikov, who also led pioneering research on *Anopheles* mosquitoes, the genus that includes malaria carriers. Kozhevnikov and colleagues regarded the advancing footprint of human society with alarm and proposed that pockets of untouched nature be protected as scientific "controls," intact benchmarks against which biologists could measure the environmental mayhem unfolding outside their boundaries. They saw ecosystems as natural places that would remain in static equilibria as long as humans could be kept out. The fitting term *zapovednik* comes from *zapoved*, a Russian Orthodox religious commandment used to regulate various aspects of daily life, including forest protection. So, at a time when Gifford Pinchot and Teddy Roosevelt were founding America's public lands system based on a credo of extensive human use, Russian scientists were proposing a network of strict biological reserves free of people.

While the zapovednik advocates' view of natural stasis has been thoroughly overturned by biologists, their notion of protecting representative ecosystems is widely embraced, and the Russian reserves themselves have performed admirably. The thickly wooded Barguzinsky Zapovednik, among the strictest protected areas in the world, averted the extermination of the sable and, to this day, provides a home for brown bears, moose, and even the unique nerpa, an endemic freshwater seal that hauls out on the Baikal shore. In a major 2009 review, the World Wildlife Fund–Russia concluded that 28 out of 101 federal zapovedniks existing at the time (now there are 103) were "pristine," a super-high standard indicating that they rated as high as possible on five criteria: (1) sufficiently large to accommodate ecosystem processes, (2) undisturbed, (3) bordered by undisturbed *surroundings*, (4) inhabited by a full complement of species that would naturally occur in the habitat, and (5) home for globally rare or otherwise important plants or animals. A majority of the 101 reserves got high, if not perfect, composite scores. Of the various criteria, the zapovedniks did best on the all-important number 4, the most direct measure among the five of ecosystem integrity.

A unique alliance of students and scientists pushed these reserves persistently over the course of a wild and violent century, even in the late 1930s, when the political perils of activism and association with the international scientific community were at their peak. The number of protected areas dipped at least twice, in both Stalin's and Khrushchev's autumn years, but the long-term trend has been upward.

In the 1990s, just when many countries around the world began to create scientifically inspired protected areas similar to the zapovedniks, Russia switched its focus to national parks with the recreational logic typical of the areas set aside in North America a hundred years earlier. Russia's first national parks were established in the 1980s, but the idea of parks for tourists really took off in the 1990s, when they increased from ten to thirty-eight. Ten more have been created since 2000, bringing the total acreage to over 38 million.

One example is the Tunkinsky National Park, southwest of Lake Baikal in central Siberia, which was created in 1991 covering the entire 2.9 million acres of the administrative region by the same name. The declaration aimed to draw tourists to a scenic valley with abundant wildlife and thermal waters, all features of Yellowstone (created in 1872) and Canada's first park, Banff (1885). Sergei Natvevich, the region's tourism official who greeted us with Mongolian ballads, has spent the last twenty years involved in the park, first as a ranger and subsequently in his job with the regional government. "I want to bring people and nature together," he said one day as he guided us alongside the Black River, a small waterway just shedding its snow and ice in mid-May in 2019.

We stopped at a Chinggis (Genghis) Khan shrine, five poles topped with shaggy horsehair. The 9-foot-tall warriors stood among spruces and newly leafing birches and poplars, wearing metal headbands and trident crests ornamented with medallions. Prayer ribbons were tied in profusion at chest level. We continued walking up the road while Natvevich

Chinggis Khan shrine in the Tunkinsky National Park of south-central Siberia.

fielded a call from the office. Presently, we encountered a baseball-sized ruddy shelduck, maybe a week or two out of the egg, peeping in panic. Natvevich, phone to ear, chased the surprisingly fleet bird around the road. He finished the call, and with both hands free and an assist from his visitors, trapped the duckling. He explained that it will grow up into a vivid orange waterbird sacred among Buddhists. Accomplished monks often assume its form after departing this life. We sought the shelduck parent for a while and then released the peeping foundling into a likely bit of soggy forest by the side of the Black River, hoping for the best. Natvevich led us to the end of the trail and a string of healing springs.

Sergey Natvevich and ruddy shelduckling.

One was for kidneys; another, thronged by dolls and toys, brought offspring. The water was so saturated with iron that it colored the creek bed orange. We filled water bottles and sipped the heady solution little by little. Natvevich downed his bottle in one go.

The generic rubric "protected areas" embraces scientifically chosen, inviolate wild areas like the Barguzinsky Zapovednik as well as parks like Tunkinsky, set up for visitors to stay awhile and revel in nature. Both are essential to big forests and the health of the entire planet. Protected areas go by many names, including reserves, preserves, conservation units, sanctuaries, management areas, national forests, state forests, reserved zones, ecological stations, biological stations, special study areas, and wildlife refuges, to name just a few. Some of these are different names for the same thing. Others vary in what they allow.

The International Union for the Conservation of Nature (IUCN) employs six categories of strictness to bring some order to this salad of nomenclature. The strictest, like the zapovedniks or Brazil's biological reserves, are category I. The only people allowed in are researchers. We verified this by trying and failing to enter the Baikalsky Zapovednik, another reserve on the lake's shore. Category II are places, such as US national parks and Tunkinsky, that encourage recreation and forbid most resource extraction. Category III is applied to areas that are usually small and that protect "monuments," which can be cultural or natural—like caves and special rock formations. Category IV is another designation for parcels that are often small, in this case crucial for saving particular species, such as the Florida panther, for which the United States has a 24,400-acre national wildlife refuge. Categories V and VI allow more human activity, within bounds that ensure that ecological processes and species are also thriving. Category V areas are more heavily modified landscapes—for example, by centuries of traditional farming—while category VI areas tend to be large natural landscapes with low-impact resource use. Examples of category VI include Peru's communal reserves and Brazil's sustainable development reserves and extractive reserves, where people fish, hunt, and collect non-timber products from the forest.

For simplicity we use the terms "park" and "protected area" interchangeably for all of these categories. They are places a society has decided to hold in common, with special rules to govern the relationship humans will have with the nonhumans; people agree to treat an area according to certain principles that moderate its exploitation. American conservationist Aldo Leopold argued that people and the biological world are all parts of a single interdependent community, whose members, human and nonhuman, need to be treated ethically in order to sustain the whole. He called this intentional relationship, characterized by respect and restraint, "a land ethic." It's not what the ecologist saw around him. Leopold invoked slavery in ancient Greece as a parallel to the way Americans of the mid-twentieth century were treating nature—as a collection of beings not yet covered by the ethics that governed our community. He proposed the land ethic to change that across America's ecosystems, public and private.

Public protected areas are the spaces where nation-states have most fully accomplished Leopold's idea of ethical relations between humans and the rest of our ecological community. While there are millions of private land stewards across the globe, few have protected forests on a large scale. One thinks immediately of the vast Patagonian lands in Chile and Argentina secured by Douglas and Kristine Tompkins. They saved a lot of land—and have eventually given most of it to the Chilean and Argentine governments for parks.

The idea of a land ethic, of course, wasn't new when it was published in Leopold's *A Sand County Almanac* (1949). For untold millennia, people have set rules for their lands and waters and have put specific zones off-limits to certain activities. Some places are reserved for burial. Some, like the *sroun* in the Momos' forest, are for the ancestors to dwell. Some areas are set aside for game animals to breed. From a hilltop in Kaska territory, John Acklack pointed to such a place, a bottomland forest of spruce, bare aspens, and willows along the Pelly River. "That's our moose farm," he said. The First Nation traditionally restricted hunting in this zone except for cases of dire emergency.

Part of the solution for the megaforests is a big expansion of forest protected areas. Multiplying public reserves is affordable and practically feasible because governments already hold most of the needed land—which is a bit surprising on the crowded planet of the twenty-first century. Where did all that land came from? By at least 10,000 years ago, humans had explored all the world's forests. They were probably in the Congo by between 60,000 and 100,000 years ago, in New Guinea by 50,000 years ago, and in Siberia by 45,000 years ago. We know that people were in North America by 30,000 years ago and were well south of the Amazon by 14,000 years ago. For thousands of years before our modern era, all forests knew the footfalls of people. Forests supplied people with food and various other needs, from caribou parkas and snowshoes to blowguns and penis gourds. People were everywhere.

Then well-armed and diseased Iberians, Cossacks, French, English, Dutch, Belgians, and others started to roam. Enabled by caravels, small ships that could easily sail into the wind, explorers saw most corners of the planet by the early 1500s. Over the course of several centuries, vast areas of the woods were dispossessed of their people and some, for a time, not occupied by new owners. In the United States, for instance, our love of mountain wilderness came into full flower at the end of the nineteenth century, so hard on the heels of the continental genocide that there was a chance for early environmentalists to thwart wholesale industrial development of the mostly depopulated landscapes. Yellowstone's establishment, in fact, came eighteen years before the US Census Bureau declared the American frontier officially defunct in 1890. The US Army was stationed in the park to discourage attacks by the last of the region's free Native Americans. Before contact, twenty-six different tribes had connections to the Yellowstone area. America's largest temperate intact forest landscapes are protected within the 16-million-acre Tongass National Forest, which was created in 1907 on the traditional territory of the Tlingit, Haida, and Tsimshian peoples.

In New Guinea, the story was different from that of the other four megaforests in ways that make protected areas both tricky and not really necessary. The Dutch, English, Germans, and Australians all came, seeking feathers, logs, gold, and oil. But in the forest interior, old ways of allocating and stewarding land persisted and were legally enshrined after PNG's 1975 independence and western New Guinea's 1969 formal takeover by Indonesia. People kept their land.

In the other four megaforests, protected areas done right can be an ethical response to history. Protected public land has prevented the industrial civilization from delivering the ecological coup de grâce to landscapes that have lost large numbers of their original people. These places have weathered the last half-millennium of economic "progress" miraculously well. Now they need protection. Modern nature conservation needs to go about this in ways that acknowledge the injustices that determined current ownership of forests and values the people who are still in them.

One of the most interesting international trends is the fusion of Indigenous guardianship and conventional protection. The Udege's embrace of the Bikin National Park is one example. A similar park was created by tribes in Colombian jungles heretofore shielded by thundering falls and a treacherous canyon on the Apaporis River. Gold mining was a growing threat. Native tribes already had legal rights to their forest under the country's robust Indigenous land law. Miners, however, could still try to entice leaders with money, splinter the community, and get at the rocks. So community leaders asked the Colombian park service to create the Yaigojé Apaporis National Park—right on top of their traditional land. That status, secured in 2008, eliminates the mining threat with no curbs on traditional uses of nature. The creative move won locals the 2014 United Nations Equator Prize, a recognition of efforts that address poverty and conservation at the same time.

Nowhere is this trend more apparent than in Canada. Rather than push out Indigenous peoples, as was done with early parks, Canadian society has put them at the center of its conservation campaign. The boreal forest is now the stage for a redemptive process involving First Nations, territories, provinces, and the federal government, with shared visions for massive-scale protected areas emerging. Indigenous Protected and Conserved Area (IPCA) principles were fleshed out by a group called the Indigenous Circle of Experts in 2018. Government then solicited proposals from First Nations. As of mid-2020, there were approximately 130 million acres of proposed IPCAs in the planning stage, including an almost continuous arc of forests through British Columbia, the Yukon, and the Northwest Territories. The first new IPCA formalized was 3.5 million acres of woods and water inhabited by bison, caribou, tundra swans, and white-fronted geese in the territory of the Dehcho Dene people in the Northwest Territories. This Edéhzhíe Protected Area, created in 2018, is double the size of Banff and will be stewarded by guardians called the Dehcho K'éhodi.

The second was a 6.5-million-acre expanse of boreal forest ecosystems called Thaidene Nëné, protected by another Dene people, the Łutsël K'é, in cooperation with the Northwest Territories and federal government. Legally protected forest now wraps around the southeastern side of the massive Great Slave Lake, whose shoreline features dramatic glacially carved promontories. Steven Nitah, who led negotiations for the First Nation, explained in a local press report, "The common denominator in all of this is the Łutsël K'é Dene First Nation and Łutsël K'é Dene law that we use to protect Thaidene Nëné already. . . . So what we're bringing to this relationship is Crown laws to operate in collaboration and unison with Indigenous Łutsël K'é Dene law."

A proposed Ross River Kaska IPCA in the Yukon would formally protect 10 million acres and give the Kaska a measure of control not seen since precolonial times. It's 64 percent of the Ross River community's traditional territory and contiguous with IPCAs planned by several other Kaska First Nations to their south. The combined proposals would protect around 40 percent of their wider 60-million-acre traditional territory, the great majority of it intact forests. The Ross River proposal cites reams of scientific data and highlights benefits for caribou, in addition to a long list of other animal and plant species of official governmental concern. The IPCA document has jargon like "boreal low bioclimatic zone" and prickly acronyms like WKA, which stands for Wildlife Key Area—the IPCA has twenty-six for moose.

But the beating heart of the proposal is the Kaska's obligation to their place. The land-use plan on which the IPCA is based is called *Gu Cho Kaka Dee*, meaning "Our Ancestors' Instructions." These instructions bring into the protected area plan historic battle sites, traplines, sacred locations, and places where traditional foods are gathered. Kaska elder Norman Sterriah has been leading the push for the IPCA. He wears a long, sparse beard and looks out through wire-rimmed glasses across the chunks of early ice riding the Pelly River's current in Ross River. He explains the IPCA in dense, sibilant Kaska, translated by Josh Barichello: "I've really been pondering over it, and, in the end, I've been thinking about the Dena Code of Ethics. It's called IPCA in the White-man way. If we live by the Dena Code, things will be good for us. . . . The creator

Aurora borealis blooms over the proposed Kaska Dena IPCA.

made this land for us. He put game on top of it for us. He made water and birds. He made everything. Long time ago Sugayeh Dena, he walked on this land. He spoke to the game. That's how connected we were. We all were that way.... The way we watch over it is through Traditional Law on this land. If the White men can see this with their sinewy eyes, maybe they can finally tell the truth when they speak."

Protected areas have proliferated worldwide over the last century, especially in the last thirty years. In 1990 many countries had none, and a mere 4 percent of the planet's land was protected. Now, nearly all nations are parties to the Convention on Biological Diversity (CBD), which set—and largely reached—a goal to protect 17 percent of land by 2020. This is a stunning global success for nature.

Protection's progress has been nothing short of astonishing in the tropics. Tom got his first look at the Amazon forest in 1965. He was initially nonplussed as he confronted a green wall with few visible animals. Then he started to notice the forests' uncountable shades of green and diversity of leaves, which seemed rarely to repeat from one tree to the next. He saw ant varieties in unimagined profusion and began distinguishing the myriad sounds employed by organisms to communicate in a setting with endlessly obstructed lines of sight. Diverse as the planet's biggest tropical forest was, and is, the Amazon biome had only two protected areas at the time. One was the Xingu Indigenous Park, set up in 1961 as a refuge for tribes in the Xingu River basin. The other was Venezuela's Canaima National Park, created in 1962 in the Orinoco basin.

Elsewhere in the tropics, parks were similarly rare. In the eastern Congo, Virunga (originally Albert) National Park protected mountain gorillas starting in 1925. Odzala-Kokoua National Park was established in the Republic of the Congo in 1935. Dutch colonials set aside a park in present-day Indonesia to protect the Javan rhino in 1921 (still hanging in there) and another to protect people—New Guinea tribes—in 1923. And that was about it.

Most of the action has come since 1990. In that year, 12 percent of tropical forests worldwide were protected. Protection jumped to 26 percent by 2015. Brazil has led the way, with 477 million acres of protected Amazonia, including 259 million acres of Indigenous territories. Overall protection of the Brazilian Amazon stood at 46 percent as of 2021. Indonesia comes in second with around a third as much protected forest, 128 million acres scattered across its many islands. Venezuela is just behind with 112 million acres. Next in the tropical rainforest category is the Democratic Republic of the Congo, with 60 million acres. Smaller countries have also made huge strides in the last couple of decades. Gabon, which is almost entirely jungle, rolled out a new system of thirteen national parks all at once in 2002, covering 11 percent of its territory.

A common misconception is that the big reserves in remote places across the globe are "paper parks," lines that exist on maps but do nothing to really protect nature. In fact, since 2000, intact forests landscapes in IUCN category I–III protected areas (the strictest kinds) were a third as likely to be fragmented or deforested as unprotected or loosely protected forests. One study found that 37 percent of the spectacular drop in Brazilian deforestation in the mid-2000s was due to protected areas. Another showed that strict protection in Brazil reduced forest loss by around 85 percent compared with unprotected forest. More flexible protected areas reduced deforestation by nearly 60 percent. That's not perfect, but it's not "paper" either. In the Congo, the only intervention that makes a statistically significant dent in the deforestation incited by an upgraded road is a protected area.

The two biggest global environmental accords in force today are the climate and biodiversity treaties, both born at the 1992 Earth Summit in Rio de Janeiro. The Global Environment Facility, a funding entity, was also created at the conference. Nature protection took off after the summit. In the previous decade, the World Bank had made some environmentally disastrous loans in the tropics, leaving development agencies generally eager to do something good for the forest. They were ready and willing to fund conservation and were subsequently joined by private philanthropists such as Intel cofounder Gordon Moore, whose foundation gave the unprecedented sum of over $450 million to Amazon protected areas between 2003 and 2021. In many tropical countries, the first generation of homegrown conservationists had emerged in the 1980s and was ready to seize the moment.

The Convention on Biological Diversity parties have now upped the protection target to 30 percent by 2030. This equals 4.3 billion additional acres and includes all types of ecosystems. We estimate that around 1.4 billion acres of core intact forest landscapes need protection, with other forests to buffer and connect them. This is a geographically monumental but quite affordable undertaking. Using a model developed by Venezuelan biologist and geographer Janeth Lessmann, we calculate that it would cost around a billion dollars annually to manage 1,120 new protected areas of around 1.25 million acres each, which is approximately the average IFL size. Lessmann's model reveals that as the size of a protected area increases, the cost of protecting each acre plummets. Ten times as much protected forest—1.25 million acres instead of 125,000 acres—can be managed at only twice the cost. According to Lessmann, the annual opportunity cost, which is the amount of income forgone from uses, such as farming, that aren't compatible with the new parks, adds another billion to the tab.

Management and opportunity costs together amount to $2 billion per year to save the unprotected IFLs. Is that a lot? Well, in 2019 humans worldwide spent $94 billion to feed animals living inside our houses. If we put a pet-meal worth of money in a jar every forty-seventh time we fed our tame animals, we could fund forest homes for trillions of plants and animals that can capture their own food. And we'd reduce the planet's future temperature and keep its water circulating. Even if we've somehow undercounted the IFL protection cost by a factor of ten, it's still cheap; the $20-billion-per-year price tag for protecting the intact forests would be just 0.002 percent of the 2019 global GDP. A bargain.

The economics are favorable but oversimplify the challenge in one crucial respect. The prospective protected areas are more than just lands for which the opportunity and management costs need to be paid in some straightforward transaction. People live there. In a world of 8 billion human inhabitants, 1.4 billion acres of new protected areas implies overlap between people and trees. By one estimate, protecting half the planet's land, a proposal espoused by eminent scientists such as E. O. Wilson, would require protecting territory already inhabited by a billion people. *Existing* protected areas are home to 276 million people. Human populations are relatively small in intact wilderness; nonetheless, forest parks and people must thrive together. The good news is that there are good working models.

A place called Mamirauá is one.

Mamirauá Sustainable Development Reserve has no gate and plenty of local *ribeirinhos* going about their ordinary lives. *Ribeirinho* roughly translates to "riverside people" and refers to populations of diverse ancestry who dwell along the Amazon's rivers and live off the land in ways that are in many respects similar to the lifestyles of the basin's Indigenous peoples. They plant manioc, squash, and beans in the dry season when Amazon waters drop more than 40 feet (12 meters) and expose beaches fertilized by Andean silt. They fish, hunt, and, in Mamirauá, operate a floating hotel, called the Uakari Lodge. The name comes from the reserve's star attraction, an uncouth monkey that's completely bald with sunburn-pink scalp and a downy white fleece covering the rest of its body. It looks like a small man who escaped from a tour group in Manaus and disguised himself in an albino gorilla suit with no mask.

When we visited in July 2017, water covered the bottom 30 feet of tree trunks, having already fallen 10 to 15 feet from its peak. The flooded forest is called *igapó*, an Old Tupi term meaning "root forest." Guides paddled us along water trails among the labyrinth of trunks. The canoes' 4-inch freeboards put our elbows a hair above caiman eye-level. Mamirauá has two of these alligator relatives, the smaller and less fearsome spectacled caiman, a favorite food for jaguars, as well as black caimans, which can grow to 15 feet long and will eat people. Several years ago a young biologist who dangled her leg from the dock of the Uakari lodge miraculously survived a trip to the bottom of the river with a black caiman, but did lose the leg.

The woods were both chatty and still—full of bird, insect, and monkey talk, but hushed enough for us to notice the kiss of paddle on water.

Pink dolphin surfaces in Mamirauá.

Our guides pointed out squirrel monkeys, sloths, trogons, spectacled caimans, slumbering nighthawks, woodcreepers, toucans, and cartoonish toucanets, all while in deliberate pursuit of the evasive bald uakari. We eventually saw a flash of white amid suddenly swaying branches as the monkey tore through the canopy.

Mamirauá was championed by Márcio Ayres, a biologist who grew up at the mouth of the Amazon in Belém. He saw his first uakari far from the jungle, in a German zoo. The young man was captivated and several years later decided to pursue PhD research in a region known to have a healthy population of the monkeys. In the course of his data collection, he convinced authorities to close a lake called Mamirauá to fishing boats, which were interfering with his study. That made Ayres unpopular, even provoking death threats. But he managed to gather his data and head back to Cambridge University to write his dissertation. When he returned to the Amazon, one of the local old-timers, Seu Joaquim, who had bridled most at Ayres's high-handed lake closure, urged Ayres to help close it again. It turned out that the biological response to protecting the lake, an arapaima breeding spot, had quickly spilled bounty into the surrounding braid of rivers, channels, and flooded forest. Now, after several years without protection, the fish population was tanking again.

Ayres and his wife, Carolina, worked with local communities to establish a federally recognized reserve, with all the villages, crops, and fishing spots inside it. The locals started managing the fish stocks with the help of a scientist brought in by Ayres. The biologist organized a periodic arapaima census, using a big net to catch all the fish and tag them, which took a week and lots of help from fishermen. The *ribeirinhos* amused themselves taking bets on how many fish there were before the nets came up. Their guesses were uncannily accurate. After repeating the exercise enough times that the workings of chance were ruled out, the fish expert started letting the *ribeirinhos* do the stock assessment their way, which took around an hour instead of a week and didn't stress out the fish. Their secret? Arapaima have a mixed breathing system, using both gills and air gulped at the surface. Because the fish surface at consistent intervals, the people who had been around these animals all their lives could easily calculate the stock just by watching the water for a while.

In the late 1990s, by collective agreement, the *ribeirinhos* started rigorously limiting their catch and only going after the really big arapaima, some of them 8-foot, 400-pound specimens. For young people, these were the leviathans of legend. For the elders, they were improbable visitors from the past. The stock increased 200 percent in the first three years of the program, which has now been replicated in locations around the Amazon.

The Mamirauá reserve gave communities the rights to enough forest and water that they could manage collective resources, like tourists and fish, judiciously, rather than competing for every last arapaima or uakari viewing spot.

Ribeirinhos have been fishing, hunting, and farming here for generations, periodically absorbing technologies such as outboard motors, radios, and modern medicine. Protecting their land and waters in sustainable development reserves is a win for the planet, ensuring that various species survive and that the global and South American climates remain more stable. Communities win security for their collective territory, and on a practical level, their relationships with fish and monkeys minimize the need for protected area staff. The community members have their own motivations to protect forests and the rivers that feed them.

These areas have proliferated thanks in part to Rita Mesquita, a Brazilian biologist who studied birds in the 1980s in Tom's forest fragments project, at the site near Manaus where scientists investigate the effects of isolating rainforest patches from the contiguous megaforest. The first in her family to attend university, she went on to earn a PhD at the University of Georgia and then returned to the Amazon, where she has lived for nearly four decades. In 2004, Mesquita left a comfortable job with the National Institute for Amazon Research to create protected areas for the state of Amazonas. "When I got to my new office there was nothing, not a single sheet of paper, not even a pencil. There was one phone balanced on top of a cubicle divider," she laughed during a video call in mid-2020 from her home in Manaus. Mesquita radiates a maternal instinct for the world, critiquing problems, incompetence, and misguided approaches with incredulous affection. She has the empathy of someone who has tried, and sometimes failed, to solve most of the problems she enumerates.

Mesquita is a force of nature; during four years in government, she expanded the state's protected areas by 13 million acres, an increase of 76 percent. Most of the new protected areas were similar to Mamirauá, serving as legal scaffolds for the *ribeirinhos*' rainforest economy. These areas have similar practical advantages to those seen in Indigenous territories. Locals continue largely traditional lifestyles in an intimate dialogue with forests and the rivers that meander through them.

When asked to single out a model protected area, Mesquita said, "There are lots of examples. The apple of my eye is the Uacari Sustainable Development Reserve, a protected area we created." It's named after the same monkey that draws tourists to Mamirauá. "You have strong social organizations inside the territory, so what happens in there is going to be what the locals want. They have a history of resisting logging in the floodplain forests. Uacari's leaders are really strong." She extolled their work on education, social services, and success in bringing forest products to market to improve living standards. The narrative is all about the social dynamics within the communities, not scientific decisions by outside resource managers. There will always be tempting opportunities for *ribeirinhos* to sell timber and overharvest game; none of these forest communities is an ecological utopia. And while the formality of protection can relieve communities of some pressure to abandon nature-based economies in favor of big, brief windfalls, the forest ultimately depends on how well the humans work together.

When we talked to Mesquita in 2020, it was two years into the most anti-nature government Brazil has had since the military dictatorship of 1964–1985. We were curious about what she saw happening to Brazil's 170 million acres of still undesignated public forest, an area the size of France. "I believe that in the future we are going to create protected areas once again," she predicted. "It's not going to happen right now. What we have to do now is take care of the system we have. I think it's most important now to show that well-managed protected areas can generate benefits for society. This is the time to create good models."

When politics do become favorable for new reserves, what places should be prioritized? That's a question we asked Adrian Forsyth, a beetle expert from Canada with four decades of Amazon conservation experience at the Smithsonian, Conservation International, and the Gordon and Betty Moore Foundation, among other organizations. "If you can go on Google Earth and find intact forest without roads, it needs to be protected. Today." The nuances of species distribution and other ecosystem characteristics are interesting, but largely beside the point, he says. "You don't need to set foot there to know it's worth it. That's what science has *already* told us." The science he refers to is the function of intactness in the large-scale biophysical health of the planet, the relative richness of anything that's whole, and the increasing scarcity of such abundance.

Chris Filardi, an American evolutionary biologist and chief program officer at Nia Tero, a group that supports Indigenous guardianship of ecosystems, concurs, sounding a more spiritual note. "It's not about an algorithm to decide what to protect. It's just about where is there space?" says Filardi. "Protected areas are about saving space so people can love it. The deeper values are what matter. They are about love and mystery as much as knowledge and knowing. Protected areas are our immune response to an unknowable future."

After several years of smashing success protecting almost inconceivably large tropical forests, Rita Mesquita realized that the state of Amazonas had tens of millions of acres of protected areas whose purpose was unknown to ordinary people. "I got to thinking about the need to communicate with society more directly about biodiversity and the value of the Amazon." Her chance came with a 2008 invitation to run a new Amazon museum to be nestled in the 25,000-acre Adolpho Ducke forest inside Manaus. She knew she couldn't get all the city dwellers out into the wild jungles, so she brought the jungle to the people.

With Mesquita and her bird-talking husband, Mario Cohn-Haft, we climbed a 138-foot tower in the middle of the forest to watch the full moon rise on a warm, pre-pandemic evening. At the top we found ourselves among a throng of young Manaus urbanites taking selfies with the rainforest as background. Cohn-Haft scanned the canopy with binoculars and pointed out a pair of blue-and-yellow macaws. Mesquita explained that this is the place where thousands of Amazon dwellers have come to be wowed by a forest that, in a sense, they didn't previously realize they lived in. "If we have any hope of maintaining our protected area system with a positive outlook and purpose, that hope resides completely within civil society," Mesquita said. "Conservation can't be an elitist thing. We have to make nature accessible to everyone."

We agree. To protect big places, our societies must also protect some small ones. Most big forests are still big because they're hard for people to get to. The abstract notion of "forest" is emotionally inert without personal contact. The park in Manaus provides that contact. It is an outpost of wild nature that gives multitudes of urban Amazonians their opportunity to be entranced by the megaforest. Smaller forests can calm the heart and dazzle the eye and make you want to breathe deep, lose track of time, and wander farther in. Doing so reminds us that there are places in this world where nature goes on and on—forest interiors like rooms leading to more rooms, far enough that there is space for unseen beings to simply be.

10

Less Roads Traveled

If a person were to be thrown out of a plane over one of Earth's inhabited continents (with a parachute), and if she should land on a spot without a road, on average she can expect to be able to walk to one in around 45 minutes. In Europe, a 15-minute stroll will get her to a road. On the chance she should alight in the Canadian boreal forest, however, she should bring provisions, because the largest roadless area there is around 450,000 square miles. That's a square 670 miles on a side, more than the area of Texas, Oklahoma, and New Mexico combined. Geometry and probability dictate that she can expect to land 112 miles from the nearest road.

And if our parachutist comes to Earth in the world's largest roadless area, she is out of luck unless she's found by helpful Amazonian people. She can expect a minimum 145-mile trek through the jungle to the nearest road. Tropical jungles, it should be said, are notoriously hard to navigate in straight lines.

These calculations are based on a 2016 map of the globe's roadless zones. Created by German researcher Pierre Ibisch and colleagues, the map displays roads in red. It assigns the same color to a 1-kilometer (0.6-mile) buffer around each road and paints the rest of the world a spectrum ranging from orange to deep indigo. The warmer colors represent lands closer to roads; the darkest blue is for the areas farthest from them. The lower forty-eight United States, southern Canada, Europe, Japan, and South Korea glow solid red with bits of orange peeking through. The five megaforests are deep, dark blue.

To get a feel for what it's like to go from a red place to a blue one, we'll land in a town called Rurrenabaque, in the Bolivian Amazon. On the eastern bank of the Beni River, Rurrenabaque has a small grid of paved streets and some dirt ones on its outskirts and also a tiny airport that upgraded from grass to potholed tarmac not long ago. There are hotels, taxis, Internet, and a bar called Moskkito, where you can order mixed drinks in Spanish, English, or Hebrew.

The town is a jumping-off point for trips to the Madidi National Park, which is, by some measures, the world's most biologically diverse protected area. The park attracts thousands of visitors annually to a zone along the Beni and a scenic tributary called the Tuichi that are easily reached in several hours by boat from Rurrenabaque. But the park has extremely remote areas as well, including the headwaters of the Madidi River itself. The isolation of the upper Madidi made it former Bolivian dictator Hugo Banzer's choice location to imprison critics, many of whom perished at the camp there. Years after the camp was closed, a 1990 scientific expedition in the area revealed such stunning biodiversity that researchers immediately called for its protection, which led to Bolivia's declaration of the park in 1995. Having always been curious about this place, called Alto Madidi, in October 2016 John and his wife, Carol, arranged to travel there with several local guides, led by a Tacana Indian named Darwin.

We caught a water taxi across the Beni and followed a gravel road several hours northwest past a patchwork of fields and forest and occasional evangelical churches. Small roadside billboards sported agency logos and titles of projects to supply water, install culverts, and foment chicken husbandry. Two hours of driving brought us to Tumupasa, a center of the Tacana community, and two more to the last town where cell phones worked, Ixiamas. From there, we veered west on an increasingly rough dirt road. Sawmills appeared periodically, as did the identical brick houses the Bolivian government gave away to raise living standards on the frontier. Built for the highlands, the houses heat up like pizza ovens in the blasting tropical sun. People mostly use them for storage and continue to reside in their old wooden dwellings.

Further on, a cart of tidy blond Mennonite children was towed by their equally well-groomed father piloting a motorbike through puddle and rut. Migrant women from the highland mining province of Potosí walked the road in full skirts and broad-brimmed sunhats. The last town on the road is El Tigre, which is what they call jaguars in the region. Established in the 1990s, El Tigre includes a tiny store, a few dirt streets, and a small collection of wooden houses.

Continuing along the road, we traversed smoldering remains of a recently burned forest. It smelled like wet campfire. Skinny remnant trees twisted in the glare while others lay charred and crisscrossed on the ground. After another hour of driving, we finally entered the cooling forest canopy and came upon a small camp where a young mother prepared food under a blue tarp. Her daughter tottered around piles of provisions—onions, potatoes, oranges, avocados, and firewood. There were filmy tents, stump furniture, and a pair of motorcycles standing by. We stopped to chat. She explained that family members were out planting pasture in newly cleared fields, places that have just blinked pink on the Global Forest Watch map.

The end of the road was a logging camp. There were no loggers but scores of logs, some 5 feet in diameter, along with milled boards stacked amid great piles of scrap wood. Vacant tents stood on platforms near a makeshift kitchen. Trash was everywhere: flashlight batteries, plastic bottles, purple and orange drink powder envelopes, junk food wrappers, toothpaste tubes, empty shampoo packets, stacked empty egg flats, and broken tent poles. "PELIGRO" ("DANGER") and a skull and crossbones were spray-painted on a lone white tree left standing in the clearing. The camp was next to a dry creek that Darwin said used to flow clear from the forested hillside.

At the far end of the camp, beyond all the tents and trash and logs, the clearing widened into an empty expanse of mud where yet more logs had

Logging camp at the end of the road near the Madidi National Park in Bolivia.

recently been collected before being trucked to the mills we passed. Tractor tires had printed the mud, providing water a place to collect and incubate mosquito larvae. We camped on the edge of the mud, chewing coca leaves and burning a little offering of cigarettes and alcohol to Pachamama, the Earth spirit, for a safe eight-day trip in the woods.

Before we started on the footpath the next morning, a motorcycle came whining out of the forest loaded with two passengers and a huge sack stuffed with game. A monkey tail flopped out the top.

During the first couple of hours on the path we saw evidence of hunters: tracks, a camp swarming with flies, feathers of a plucked jungle gamebird—a curassow, a turtle shell, and discarded soda bottles. Later, just forest. Creeks ran in different shades, some iron-orange, others black tea, still others clear. Darwin drank from them all. Then we smelled the funk of white-lipped peccaries, a native species of wild pig. Actually, we heard them first: reports like gunfire as hard-shelled palm fruits cracked in their mandibles. Following Darwin, we crept up on the troop of around a hundred animals feeding and loitering in a gully. Catching our scent, the pigs spooked and stampeded deeper into the forest.

Farther along the path, now far from the road, twenty black spider monkeys swung through the branches. These primates know that they are meat and keep their distance. Seeing the spider monkeys at all is a sign we were getting into untrammeled territory. They moved hand over hand over prehensile tail, a form of locomotion called brachiating. Spider monkeys and their Atlantic Forest relative, the *muriqui*, are the only primates in the Americas who brachiate. The rest run along the tops of branches. The purest brachiators are Asian gibbons, but all apes—gorillas, chimps, bonobos, and orangutans—do some version of it, including those human primates you see in playgrounds, brachiating while they are still light and strong enough to use the monkey bars.

The light faded as we walked the last mile to the Madidi River, arriving on the spot where Banzer's prison camp used to be. The next day we wandered upriver to a stand of balsa trees. We dragged out as many fallen trunks as we could and then downed several more with machetes. At Darwin's direction, we lashed the logs with strips of their own inner bark, making a craft around 20 feet long and 5 feet wide, with a small raised platform in the center.

We cast off the next morning, floating the horseshoe meanders of jade-green water, seeing no other people. Darwin said that an average of one group of visitors comes out here each year. The second evening we beached our raft in the evening at a prime fishing spot, a deep pool where we threw handlines while a 6-foot giant otter sat on a log crunching a plate-sized specimen of a prized fish called *pacu* (*Colossoma macropomum*). A huffing from the bush told us a jaguar was watching us watch the otter eating the fish. In the morning the huffing was closer. We tumbled out of tents and dashed down the beach, around bushes, and found ourselves, for a

Fresh jaguar tracks on the bank of the Madidi River.

fleeting moment before she veered into the trees, face to face with the Western Hemisphere's biggest feline, *el tigre*.

This is roadlessness, the deep blue on Ibisch's map.

The first big road in the Brazilian Amazon was, like the Bolivian town we just visited, named after the circumspect jungle cat. Construction began in 1958 on the *Estrada da Onça* (Jaguar Highway) to connect the capital, Brasília, to Belém, at the mouth of the great river. Several factors impelled its construction. One was Brazil's own manifest destiny doctrine, articulated by President Getúlio Vargas in the 1930s, known as the "Westward March," though the Amazon was more north than west for the marchers. Brazil's military establishment worried about the country's jungle borders. The generals wanted roads so they could impose order and have control in the wild north. At least as strong as these patriotic geopolitical ideas was pressure to exploit land and timber in a frontier free from the order and control the generals envisioned. The Jaguar Highway, now simply referred to as the Belém-Brasília, was finished in 1960.

Deforestation and lawlessness ensued. Land speculators, gold miners, and settlers overran the region and its Indigenous peoples. When Tom arrived in 1965, the chaos seemed to have taken Belém residents by surprise; they had conceived of the road as a way to get to Brasília and back, not as the catalyst for ravaging the space in between. Violence reached such a pitch that in the 1970s, the Brazilian military had to declare a state of emergency. Their vision of an orderly Westward March was a bust, but the generals pressed on with a propaganda campaign and incentives for Brazilians to occupy the forest.

Overall, 95 percent of Amazon deforestation has taken place within 3 miles of roads or immediately adjacent to one of the region's major navigable rivers. The Belém-Brasília was the beginning of a road-building program that later included better-known projects, such as the Trans-Amazon Highway, which spawned fishbone patterns of lateral roads probing into the forest. More recently, the region's governments and development banks have collaborated on a grand regional infrastructure program (under the

Spanish acronyms IIRSA and later COSIPLAN), which includes roads that would penetrate the deepest jungles of the Amazon. Many of the projects, fortunately, remain unbuilt, but the initiative continues.

In the Congo, deforestation has not yet metastasized from roads with Amazonian vigor, but it does happen. Immediately around new Congo blacktop, 91 percent of the forest is shorn. Nearly 20 percent is razed 1.2 miles (2 kilometers) on either side, and even 6 miles (10 kilometers) away, forest patches are lost. Those are averages. Roads are at their most devastating in intact forests, places being opened for the first time. Roads' biggest impact in Africa, however, is invisible to the satellites that provide deforestation data. It's hunting. Roads spread a plague of defaunation as people use them to move in and harvest anything that moves, big animals first, and export the meat to urban markets.

The danger is not lost on forest elephants. In one western Congo study, researchers fitted twenty-eight of them with radio tracking devices and watched the animals move around for a year, noting whenever one of the animals crossed a road. There were roads both inside and outside protected areas. Four fortunates among the elephants had no roads within their ranges. Of the remaining twenty-four, seventeen crossed roads within protected areas, some routinely. The tracking gadgets allowed scientists to see how fast the elephants were moving and revealed no particular haste when the animals were crossing "protected" roads. Only one elephant, a cow called Mouadje, with calf in tow, crossed a road outside a protected area. She did so three times, uncannily finding the single largest gap between the 134 villages along the road—and moving at a dead run. On a normal day, she traveled an average of around a mile. On a road-crossing day, her travel distance exploded to 15 miles.

Elephants know that roads bring death, especially outside protected areas. A mere three crossings of unprotected roads in a cumulative 28.5 years of elephant-time show that roads are like fences, only more confining because the animals know better than to even approach them. The probability of stepping in elephant dung in a western Congo forest skyrockets once you're 15 miles or more from a major road.

In the boreal, roads block streams, lead to more fire, and threaten

Logging road outside Nouabalé-Ndoki National Park.

game animals like caribou. Elizabeth Dabney, executive director at the Northern Environmental Center in Fairbanks, Alaska, cites the example of the proposed 220-mile Ambler Mining District Industrial Access Road. The route, supported by the state government, would cross the southern Brooks Range, traversing Gates of the Arctic National Park. The idea is to ease transportation to a series of marginally profitable mines. Eleven native villages oppose the road, in part because the communities depend on the Western Arctic Caribou Herd for protein and cultural well-being. Caribou, like elephants, commonly refuse to cross roads. The Ambler project would obstruct the animals' migration from summer habitat to calving grounds. The animals' normal life cycle would be interrupted, and the survivors would be effectively penned, becoming easy quarry for hunters driving in from cities and towns. The villagers, in turn, would lose the herds on which their existence has depended for thousands of years.

Roadlessness is central to the survival of the megaforests, both tropical and boreal, but few countries that we're aware of have explicit roadless policies. The United States is a partial exception. The practice of protecting roadless areas in the United States is nearing its hundredth birthday. It started with Aldo Leopold, who went to work for the US Forest Service straight out of Yale Forestry School in 1909. Over his first decade overseeing forests in New Mexico and Arizona, Leopold watched the landscapes falling apart due to roads, grazing, wolf eradication, and exploding deer numbers. Streams disappeared, hunks of hillsides calved into valleys, formerly open ponderosa stands filled with brush, and native trout vanished. Leopold, although then still a gung ho disciple of the development-minded Forest Service founder Gifford Pinchot, looked for a place he could establish a roadless wilderness. He started with a list of six in 1920, but roads wormed into five of them by the time he got his chance. In 1924 he created the country's first wilderness area, in the Gila National Forest in southwestern New Mexico.

Bob Marshall, a wilderness champion and explorer of remote Alaska, took the roadless baton from Leopold and ran with it. Shortly before his early passing in 1939, Marshall restricted roads on 14 million acres of the national forests.

The Wilderness Act of 1964 codified these ideas, requiring that the government identify national forestlands that merit protection from industry. Part of this process was to determine which areas were still road free. Reviews were completed in 1967, 1977, and 1998. After the last, Bill Clinton's Forest Service chief Mike Dombeck drafted a policy to protect roadless areas—not just some of them, as had been done in previous reviews—but all of them.

In January of 2001, ten days before the inauguration of George W. Bush, the Forest Service banned the construction of new roads anywhere within the 191-million-acre national forest system where they did not yet exist. These 58.5 million acres were in addition to wilderness areas already set aside from logging, mining, and motorized recreation. The roadless policy included more than half of the Tongass National Forest, 16 million acres of rainy, mossy woods with giant hemlocks and cedars on Alaska's southeast coast.

Clinton and Dombeck reasoned that the simplest and most reliable way to protect an increasingly fragmented forest landscape was to keep roads out. Private forests in the United States were getting chopped to pieces. The proportion of private forests that were smaller than 50 acres had doubled since the late 1970s. And most of the public forests were saturated with roads. The still-roadless places had healthy forests that benefited over 80 percent of the birds, mammals, and amphibians that Forest Service biologists categorized as "sensitive." In addition, these dwindling wild spaces were in high demand among people who wanted to hike, camp, boat, and generally commune with nature, a group that was growing fast.

The regulation also cited a practical consideration, namely, a lack of federal funds to maintain the *existing* 386,000 miles of national forests roads. This is eight times the length of the interstate highway system. Congress had been supplying the agency with, on average, only 20 percent of needed maintenance money. Most Forest Service roads were in substandard condition, eroding into creeks and causing other havoc. What's more, the as-yet unroaded territory was in largely challenging topography, where roads would be costly to build and maintain.

The Tongass came back on the chopping block in 2019 when the Forest Service reversed course, exempting national forests in Alaska from the roadless policy. That means that roads funded with tax dollars could penetrate our last intact temperate forests. On his first day in office, President Biden pulled an immediate U-turn, ordering a review of the previous administration's move to open Alaska's roadless forests. During an early 2020 video call, Meredith Trainor, of the Southeast Alaska Conservation Council, explained that Tongass timber roads are just one more example of lose-lose infrastructure. "In the last two years, logging has been 0.7 percent—not 7 percent—*point* seven percent, of the economy of our region, both by jobs and earnings," she said. "Life-long Alaskans have this nostalgia for the good old days of logging and logging jobs in southeast Alaska and are willing to go to very extreme lengths to resuscitate this dying industry."

The United States can contribute to protecting the world's megaforests by keeping our portion—over 67 million acres in Alaska, including the Tongass—as roadless as it is today. As former Forest Service chief Dombeck put it in a 2018 *New York Times* op-ed, "The roadless rule is an intensely conservative regulation. It saves taxpayers money and keeps our few remaining wild remote public lands intact."

Dombeck's words about the marriage of conservation and conservative spending hold true across the world's megaforests. Destructive forest roads are usually a huge waste of money. First of all, remote forest roads are expensive to build. They're usually far from sources of construction materials, such as gravel, steel, cement, and asphalt. Rain in rainforests and snow in boreal forests accumulates in rivers and swamps that require bridges and detours, causeways and culverts. Sometimes forests are intact because of treacherous terrain, like the near-vertical Andean-Amazon slopes in Bolivia, Peru, and Ecuador. In the 1990s, Bolivia decommissioned a road called the "Highway of Death." So narrow was this road, and so profound the abyss it skirted, that vehicles would switch from motoring on the right side to the left so that the driver closer to the edge could peer out his side window to keep from going off it. Due to rain, geology, and slope, the new jungle access route cost five to ten times as much as flat tarmac did at the time.

In New Guinea, home to the other mountainous tropical megaforest, roads must negotiate elaborate folds and ridges that are challenging even on foot. Walking is how people used to get into the heart of the Tambrauw Mountains until a road clawed its way there in 2007. Most bits of the island that aren't steep are boggy.

Remote roads are also costly to maintain. In the Republic of the Congo, a 240-mile ribbon of asphalt was built in 2009 by Chinese contractors in order to facilitate movement between the capital, Brazzaville, and the country's forested north. Ten years after construction, it was so extensively and deeply potholed that some segments had become more hole than not. In cross section, the remaining pavement resembled a delicate layer of chocolate frosting spread thinly over porous cake. Men brandished machetes at bushes that had crept over drainage ditches and forced traffic toward the centerline. The pavement regularly narrowed to an isthmus too slender to fit two wheels. Minibuses and logging trucks toiled slowly around these ridges. Heavy trucks were eventually banned for a time during 2019. Enterprising locals offered strings of cooked grasshoppers and leaf-packets of steamed manioc to the crawling traffic. Some people had even carved through lanes into their front yards, where more merchandise was offered.

Remote forest roads are uneconomical for a further reason—not enough people use them. John's own experience analyzing jungle roads has shown that it takes around 300 cars per day to economically justify paving. For example, blacktop was proposed back in the late 1990s for the dirt road we described running through Tumupasa and Ixiamas, just north of the Madidi National Park. But the road had less than a dozen users a day, and a loss of $25 million was predicted in a 1999 study. The finding was affirmed by another report several years later. The road remained a dirt track for over a decade. In 2011, although nothing much had changed with the nearly nonexistent traffic, the World Bank loaned Bolivia $129 million to pave this route.

Which makes one wonder.

The World Bank headquarters is housed in a gleaming edifice two blocks from the White House, at 18th Street and Pennsylvania Avenue in Washington, DC. The mirror windows on its front reflect the trees in the park across the street. The building is chock-full of economists with PhDs from the best universities in the world. They make decisions like the one to finance the Bolivian road, which menaces Madidi National Park. The lack of economic feasibility could be established with a few calculations on the back of a napkin at the bank's excellent subterranean cafeteria. Why do smart people dump money into the most destructive and unprofitable public works in the world?

For starters, the vast preponderance of road funds come from urban taxpayers who have never seen an imperiled forest and will never drive on the unnecessary roads. Loan repayment and maintenance costs come later, so part of the bill is shifted in time as well as space, from today's voters to those who will be electing governments a decade or two hence. The people who make money from remote forest roads don't pay for them.

To be more specific, the roads confer rewards on several groups. First are politicians involved in infrastructure corruption. Four recent Peruvian presidents have faced prison time for taking bribes from Odebrecht, a Brazilian construction giant that builds roads, dams, and other public works, including the $1.5-billion furrow of forest demolition known as the Inter-Oceanic Highway in Peru, which unleashed uncontrolled gold mining in the jungle. Odebrecht admitted to bribing officials of twelve different countries with $788 million. Crooked officials pitch the ceremonial first shovelful of dirt and get their money immediately. If they had to wait to see how useful the road proved to be before collecting kickbacks, authorities might choose better projects.

When it comes to banks, the explanation for white elephant infrastructure projects can be found in a situation that's called the principal-agent dynamic in game theory. Financial agents, who make investments with

Roads appear and trees vanish in the western Amazon. *(©Sebastião Salgado)*

other people's money, are paid for simply making the transaction happen, no matter how it turns out. The principals whose money is invested benefit only when the investments turn out well and incur losses when they do poorly.

The main lenders for tropical forest highways have historically been multicountry development banks such as the World Bank, the Inter-American Development Bank, the African Development Bank, and the Development Bank of Latin America. These agencies get infusions of capital from various member countries to fight poverty and boost economic growth. They all, it should be said, have evolved their thinking toward more sustainable investment decisions. Environmentally destructive road projects can still find support, however, from the newly dominant China Development Bank and the Export-Import Bank of China, which lend and invest under the banner of the Belt and Road Initiative, a global Chinese infrastructure extravaganza.

As a group, all these banks are more likely to make financially bad road loans than are commercial lenders. They use money belonging to a multitude of anonymous and unaware principals: taxpayers. Commercial banks' principals are depositors and investors who keep an eye on their money and expect market-rate returns. When it comes to projects paid for by the development banks, the agents—staff and managers and government officials—are judged mostly on getting deals done, especially big ones, somewhat independent of what the return may be for the principals.

Forest roads also benefit loggers, cattle ranchers, soybean planters, cocoa famers, oil palm growers, small farmers, and miners. That's a lot of categories but their overall numbers are small, thousands perhaps, relative to the tens of millions of people paying for each road, most of whom are unaware that they're doing it.

As a result, this infrastructure, which often makes neither economic nor environmental sense, is a chronic threat to forests. Confronting it requires good analysis, media exposure, strong courts, and solid government agencies and regulations. Studies on economic and environmental aspects of road choices are fundamental. For example, a 2019 study of 7,500 miles of proposed roads in the Amazon found that 45 percent could be discarded on economic grounds, even before considering impacts on the forest. Each dollar of transportation benefits these bad projects promised would cost more than a dollar in construction and maintenance. The team, from several Amazon basin countries and led by the Conservation Strategy Fund and the Amazon Environmental Research Institute, further winnowed the list by considering what they called "socio-environmental risk." These risks were incurred, for example, by putting roads too close to uncontacted Indigenous groups or stoking deforestation. The team concluded that the best 10 percent of projects could be built with minimal disruption to forests while delivering 77 percent of the total economic upside available.

Similar analyses can support better road choices in other tropical forests. A 2015 World Bank study on the Congo, for instance, applies various criteria to judge projects' merits across the basin and shows that road construction in the remote jungles of the central DRC causes greater percentage increases in deforestation than do projects in more accessible sites. A global study mirrored the Amazon results, finding that roadless forests can be saved with minimal economic sacrifice. The best land for agriculture, and thus the land with the most genuine need for roads, lies outside of the wild forests.

Colombian road planners use a system called Tremarctos, named after the Andean spectacled bear. Developed primarily by the environmental group Conservation International, it allows officials to quickly put a candidate road onto digital maps that display the location of protected areas, Indigenous lands, vulnerable species, and high-fire risk, among other factors. Tools like these, with layers showing intact forest landscapes and primary forests, should be made accessible to every environmental bureaucrat and transportation planner in the world. A proposed road that would reduce the size of an IFL could automatically prompt planners to consider alternate routes or other modes of transportation.

Until 2000, these sorts of technical approaches were rare. Conservationists tended to confront the threat of large infrastructure projects with either resignation or rage. Few environmentalists had the expertise to wade into these debates on a technical level. Now traditional activism is complemented by the work of a new generation of independent economists, biologists, and geographers training their skills, models, and maps on the ample bull's-eye of unnecessary infrastructure spending.

A good example is a Brazilian analyst named Leonardo Fleck. Fleck grew up in southern Brazil and fell in love with the Amazon during a research year in the Mamirauá Sustainable Development Reserve. His analytical persona is skeptical and attentive to the smallest details. In 2007 he worked for the Brazilian office of the nonprofit Conservation Strategy Fund, burrowing into thickets of files on the planned reconstruction of 240 miles of federal highway in the heart of the Brazilian Amazon. The proposal, to link Manaus to the national road network, was arguably the single biggest threat to the Amazon in the first decade of this century. Fleck calculated that the $275-million project would fuel the deforestation of 11.4 million acres over twenty-five years and rack up $150 million in net economic losses. He also tallied the economic value of carbon emissions and other quantifiable environmental damage, which drove the road's projected economic cost above $1 billion.

Fleck explained his findings to anyone who would listen. Many did, including state and federal environmental agencies, local governments, journalists, and members of Brazil's Congress. His numbers convinced people that the road had been falsely framed as a choice between development and nature. A sophisticated discussion emerged on responsible public spending, transportation alternatives, and budgeting for mitigation costs. Good sense emboldened people, including then Environment Minister Carlos Minc, to oppose the plan, sending it into hibernation for ten years, during which time protected areas had a chance to expand in the region. The project has now reemerged, which is unfortunate, but the damage will be far less than if the road had been paved a decade ago.

Stable rules and agencies help thwart bad investments. Environmental impact assessments (EIAs) were almost nonexistent in the tropics before the 1990s, and now they are almost universal. Most embody the approach of the first EIA law, the US National Environmental Policy Act (NEPA), signed by President Nixon in 1969. In practice, these documents are often thousands of pages of soporific cut-and-paste material, with no genuine consideration given to the option of *not* building a particular road, bridge, or canal. Even so, developers abhor them because the preparation and review of these cordilleras of paper slow investments.

In our experience, despite the pro forma spirit in which they are often done, EIAs give citizens some purchase. They are tools with which well-prepared lawyers, activists, and scientists can ratchet forest road decisions in the general direction of environmental and social sanity. They involve public hearings and provide something to scrutinize and critique—something on which to improve. For example, in the mid-1990s, in the Brazilian state of Bahia, a road was proposed that would impact a forest that was, for a time, the world record holder for tree diversity. Local activists used the project's EIA as leverage to convince the state government to create the Conduru State Park, encompassing forests threatened by the blacktop.

Economic feasibility criteria can also help guide public money to better projects. While most countries' finance ministries do screen government projects for economic merit, the requirements are more flexible and less widespread than EIAs, for which there are international standards and professional associations.

For countries with expanding timber economies, keeping roads and forests separate is challenging. Still, logging need not spell doom for the forest. Abandoned logging roads in the Congo melt back into forest. Closing logging roads after harvest is a cheap and commonsense policy that minimizes poachers' access to wildlife and reduces governments' policing costs. Rather than turning the main logging arteries into all-purpose routes and creating new mill towns, governments can mandate that wood processing be centralized as close as possible to already thriving towns.

The Wildlife Conservation Society's coordinator for Russia and Northeast Asia, Jonathan Slaght, told a story about one logger who started closing his own roads to save wildlife. Slaght is the author of *Owls of the Eastern Ice*, in which he describes his adventures finding, catching, and releasing members of the world's largest owl species, Blakiston's fish owl, for scientific purposes. In the winter of 2006, he was camped out in the Russian Far East searching for owls. He awoke one morning to the sound of heavy machinery. Clambering out of his tent, he encountered a man moving a big pile of earth to block a road that was Slaght's only route back to civilization. The owl expert explained that the logger, Aleksandr Shulikin, is an avid fisherman and hunter from a local village of 800 people. "His company is putting in these roads and he personally sees the detrimental impact," Slaght said in an interview. Logging roads had proliferated in the region, from 95 miles of total length in 1984 to 1,800 in 2005. The deer, wild pigs, and salmon were disappearing. So, Shulikin had started to block the roads that poachers were using to get into the forest's depths.

Years later, doctoral thesis complete, Slaght began replicating this approach with the largest logging company in the Russian Far East.

They've closed over a dozen roads since 2015, creating 125,000 acres of de facto protected areas, safe from all but the most intrepid of hunters. Slaght said, "If someone wants to walk into the forest, shoot a deer, and walk out with it on his back, I'm okay with that. It's the people driving in there with pickup trucks, shooting anything with eyeshine and loading their truck with carcasses, that's who should be kept out of the woods.... It works. It would just be nice to see it on a larger scale." Large enough for tigers, for example, which have home ranges of over 300,000 acres. Controlling roads, Slaght asserts, is the single-most important thing that can be done for tigers outside protected areas. In 2020, WCS was studying policy options to get road closing into Russia's Forest Code.

It may seem a harsh position to oppose roads that could give remote forest communities easier access to the rest of the world. Officials in government, banks, and the communities themselves regularly characterize road access as a right akin to breathable air and freedom of expression. While access is important, the unconditional right to a community driveway has no ethical basis. When the Brazilian Leonardo Fleck studied the proposed paving of BR-319 in Amazonas, there were around 400 to 500 households along its 240 miles. Building the road for their sake would cost more than $500,000 per family. Public money always has a host of competing and urgent uses to which it can be put, including education, health, collective security, and environmental protection. Roading remote forests for small groups of people is environmentally ruinous and usually an unjustifiable use of funds.

Governments can help isolated forest communities thrive without building roads. They can, for instance, subsidize air transportation, as is done for food in roadless parts of Canada and passenger travel among cities in the Brazilian Amazon. Further, satellite Internet can help transfer services like education, health care, and banking to remote settlements, sparing villagers epic walks or weeklong boat rides to jungle cities where they pick up social security checks, go to school, or have routine health and dentistry work done. Thanks to a new antenna and bank of solar panels in the West Papua village of Ayapokiar, where we met Fince Momo and her niece Anastasia, the village can get messages to the city without climbing into a vehicle. River ports and navigation infrastructure can also be improved. An Ecuadorean company, Kara Solar, and the Achuar Indigenous people are developing solar boats. This innovation delinks mobility from high-priced fuels and makes roads redundant.

Roads have been around for a long time. Thousands of years ago there were roads in India, Persia, and the Middle East. Then of course there were the famous Roman highways that accelerated military expeditions. Thanks to the vulcanization of rubber, over the last century roads have become the dominant form of infrastructure connecting parts of the human world, and most roadless places have been breached. Roads are the largest of our artifacts, like very large initials carved in the smooth bark of a beech tree. When it comes to our great forests, the urge to carve must be resisted. Some roads are simply an expensive way to accelerate the forest dismemberment. To protect ecological integrity, governments and their citizens must begin to value roadlessness. Big forests are another sort of infrastructure, one that moves us through time, not space. They are our survival infrastructure for getting from now to a livable later world.

11

Making Nature

A Central American rainforest started reclaiming the pyramids of El Mirador two millennia ago. It sent majestic mahoganies and cedars skyward to tower over a canopy of chewing gum trees and an understory of allspice and fanlike *xate* palms. Forest fragrance and serpentine roots wrapped the right angles, insistent vines climbed walls, and seedlings cracked the pavements of an abandoned city. Jaguars, parrots, pumas, spider monkeys, howlers, and deer all returned. The forest itself crept, creature-like, to retake its home. It threw a green blanket over all the history of war, power struggles, and day-to-day urban life. Trees sealed El Mirador's secrets and metabolized ecological possibility out of every joint and fissure of its stonework.

There are few spectacles of natural resilience quite like a lush rainforest growing straight out of structures people built to last forever. Begun 2,500 years ago, centuries before the famous Classic Maya cities such as Tikal and Palenque, El Mirador features three massive pyramid complexes, dozens of other structures, and giant stone tablets, called stelas, carved with pictures and hieroglyphs. Well into the 1990s, before archaeological excavation began in earnest, forest draped the ancient architecture up to the apex of the 236-foot La Danta pyramid. Trees still cover most of it. Howler and spider monkey bands skirmish in the canopy. Toucans look on, and scarlet macaw pairs arrow across the sky in the precise formation of permanent monogamists.

The forest, according to Richard Hansen, who has led excavations at El Mirador, appears to have repossessed this site due in part to the Maya's improvident use of trees. The city lies in a landscape of boggy low spots called *bajos* surrounded by drier zones with tall forests. The *bajos* are full of fertile mud, while the higher ground has a mere dusting of unproductive clay soil atop limestone rock. Today, farmers struggle to grow a few ears of corn in this ecosystem. The Maya farmers at El Mirador, however, hauled fecund mud from the *bajos* onto garden terraces and fed a metropolis. It worked for hundreds of years.

And it might have continued had Preclassic Maya not been so partial to stucco. They smoothed walls, floors, and streets with this cooked limestone plaster, which looked terrific painted. To make stucco, they had to stoke kilns with lots of green wood. Logging loosened infertile clay, which washed into the *bajos* and covered the fertile mud with an impenetrable whitish cap up to 10 feet thick. Eventually, too much deforestation spelled the end of the mud and of the city's food supply. Around 2,000 years ago, the forest started closing in around El Mirador, climbing its pyramid steps, filling the terraces, and darting roots into the walls.

This story is a cautionary tale that forest abuse can topple civilizations, the prospect we face today. The forest at El Mirador proclaims another, more uplifting truth: forests grow back as long as soil and water, sun, pollinators, and seed dispersers are present. The once-busy metropolis at El Mirador sits today at the core of the biggest intact forest landscape in Central America, known as the Maya Forest. Later Classic Maya sites are similarly nestled in jungles. There's something reassuring about these forests' show of patient vitality.

Avoiding forest loss in the first place, however, is monumentally easier and cheaper than getting a forest back. Economists Jonah Busch and Jens Englemann looked at the feasibility of financing tropical forest conservation versus recovery through payments for their carbon storage. They calculated that seven to ten times as much forest could be affordably conserved than regrown. Natural forests don't require any planting or care and provide all their carbon storage, biodiversity, and other benefits without delay. The regrowing forest can take generations to mature.

Furthermore, regrowing trees often fails to bring back the species that were once among them. For example, the wolves, mountain lions, and elk that were formerly commonplace in the eastern United States have yet to return long after forests have. Woods regrown in the eastern Amazon will often regain a sum of species equal to their pre-deforestation tally after just twenty years of recovery, but the species will be different from those that inhabited the original forest. Human cultures that lived in intact forests are even less likely to return. Reforestation is not a viable backup plan for saving the megaforests. It is, however, a non-optional complement.

Reforesting is needed because around half of the 900-plus billion metric tons of animate carbon stored in preindustrial ecosystems has already been loaned to the atmosphere. Some of it must be brought back into the biosphere to make the math of climate stability work. The Bonn Challenge, launched in 2011 by Germany and the International Union for the Conservation of Nature (IUCN), called for restoring 375 million acres (150 million hectares) of forest globally by 2020. By 2030, the goal is 850 million acres (350 million hectares), approximately twice the size of Alaska. IUCN runs a sort of information clearinghouse, cheerleading, and technical support center for the fifty-eight governments that signed on to the Bonn Challenge. Thirteen countries reported progress to IUCN in 2018, the most recent update available due to the coronavirus pandemic. They had delivered 56 percent of their aggregate 2020 pledge.

What they have pledged to do shouldn't really be called reforestation, according to IUCN staffers Swati Hingorani and Andriana Vidal. The correct term is "forest landscape restoration" (FLR). Restoring a forest landscape, ecologically speaking, is about more than planting trees. FLR can be accomplished, for instance, by intensifying agriculture on one piece of land to ease pressure on another where trees are growing back on their own. Another recognized approach is to restore species that play important ecological roles, like dispersing seeds, carrying nutrients from one place to another, or controlling the populations of other animals that may overgraze tree seedlings. A glowing example of FLR might be revival of salmon runs, which link ocean and forest food chains by bringing biomass from the sea into the depths of northern forests of the Russian Far East and Pacific coast of North America. They feeds bears, giant fish owls, and trees. Along British Columbia salmon streams, 40 to 80 percent of nitrogen in plants has been found to have oceanic origins.

The ample definition of FLR gives Bonn Challenge member countries a big menu of options to fulfill their commitments. Too big, say some experts. A group of researchers led by Simon Lewis, a scientist at the University of Leeds, calculated in 2019 that a regrown natural forest stores forty-two times as much carbon as the same area dedicated to short-rotation plantation forestry. They critique the Bonn Challenge for making no distinction between these two approaches, noting that 45 percent of participating countries' plans involves tree monoculture. In the interest of the climate and biodiversity, FLR should exclude tree farms, Lewis and colleagues say, and encourage as much natural forest recovery as possible.

Where there's healthy forest in the environs, the best choice is to stand back and let nature do the work. For example, one February 2020 morning in West Papua, we stood under the closed canopy, listening to the whoosh of hornbill wingbeats above the treetops. We were fully oblivious that this spot was a former farm plot until our host, Fince Momo, started telling stories. From the 1960s through the 1990s, her mother and aunts planted sweet potatoes, corn, and taro here. Then, about twenty-five years before our visit, they moved their food garden to Momo land closer to the village. We looked around more carefully and saw that the trees were, indeed, thinner here than in the surrounding woods. There were red pandanus trees we hadn't noticed, each with a fan of long leaves atop a bare stalk. It's a native plant people cultivate for making mats and for the fruit's antiviral properties. A concentration of *Pandanus conoideus* in the woods is a sign of former human occupation. Shortly after our visit, pandanus demand spiked in Papuan cities as the coronavirus arrived.

Used carefully, the forest that became garden readily reforests itself, and has done so for millennia. When the Momos shifted their crops elsewhere, the soil was preloaded with wild seed. Cassowaries dropped cakes

West Papua forest regrowing where the Momos once gardened. Pandanus tree at upper left.

of fertilizer loaded with yet more seeds. Fern spores drifted in, insects, millipedes, and frogs wandered along when the time was right, and the forest rebuilt itself. A magnificent bird of paradise staked out a display ground and fastidiously cleared it. Tree kangaroos returned to the spot, and wild pigs nose-tilled the ground that humans had left fallow. The Momos' old garden could find its way back to forest because all its building blocks were present.

Natural regeneration works especially well where crops have not grown at all. Immense tropical areas have been logged very lightly, often with just a single tree removed from each acre. The woods can reclaim ecological integrity if people simply retire logging roads and preclude future harvests or require low-impact extraction techniques. David Harris, from the Royal Botanical Garden in Edinburgh, has studied the Congo's flora for over two decades. He contends that an ordinary person will walk across most former logging roads in the Central African forest without noticing. The forest isn't unscathed, but, left alone, it can recover within a decade or so.

Most of the forest acreage that could be restored is, unsurprisingly, in regions that have lost most of their tree cover—the ex-megaforests. A 2019 study employed three criteria—cost, biodiversity, and carbon storage potential—to come up with tropical forest restoration priorities. They include Brazil's Atlantic Forest, the once-dense West African forests, mainland South and Southeast Asia, and the Indonesian islands of Java, Sumatra, and Kalimantan (Borneo). The study identifies one conspicuous opportunity to heal and expand a current megaforest: the southeastern Amazon. Along the Belém-Brasília Highway and a couple of other federal roads, extensive areas have been deforested, farmed, or grazed and left with a scruffy pelt of overgrown pasture. An estimated 25 to 37 million acres are classified as "degraded and under-utilized." England could fit in this damaged plot.

Paulo Moutinho, a national environmental leader and one of the fathers of REDD+, is also a competitive cyclist who, in 2017, inhaled dust and smoke on the Trans-Amazon Highway for 700 miles, pedaling to publicize assaults on the forest. His commitment and creativity are boundless in the quest for models of sustainable human existence in the Amazon. Wiry and goateed, Moutinho has an elfin twinkle in his eyes, but isn't immune to the melancholy that comes from watching Brazil lose 220 million acres of natural ecosystems in the last thirty-five years. Most of Moutinho's efforts have focused on reversing deforestation in southeastern Amazonia.

Bringing back these forests would be a boon to the whole Amazon. The basin's "flying rivers," fed by forest evapotranspiration, are turned to trickles by deforestation in the southeast. Moutinho says that the loss of a single tree with a 66-foot canopy robs the vapor current of between 130 and 290 gallons daily. As a result, less rain falls farther west, including at Brazil's far western frontier over 1,000 miles away, where Tamasaimpa's people, the Marubo, live.

"The game gets harder to follow because the ground is dry, and we make too much noise walking on the leaves. You walk along with your feet going 'chh, CHH, chh, CHH.'" Tamasaimpa comically exaggerated the indiscreet hunter's crunching footsteps. "My grandfather said that the land was angry because of White people cutting down the forest for their ranches to the South." In the territory of the meat-loving Marubos, fish used to live a charmed life. The tribe looks askance at fishing-oriented cultures, but now depends on angling to supplement the hunt.

As sensible and important as restoring the eastern Amazon is, Moutinho cautions that fully recovering a blanket of lush natural forest—with its original carbon stock—is like pushing rocks uphill. "Reforestation is not likely to happen on a massive scale because for landowners it's like throwing money away." The landowners he's talking about are people who settled along roadways punched into the forest over the last fifty years. They can't be expected to throw away a lot of money.

In the best-case scenario, where the forest regenerates on autopilot, the cost of recovery is around $150 to $250 per acre to fence off land so that cows don't browse the seedlings. Unfortunately, most degraded land has baked in the tropical sun and hardened to a concrete-like consistency under the meanderings of half-ton cows, whose hoof-falls max out at around 290 pounds per square inch. A human male puts an average of 8 pounds per square inch of pressure on the earth. To reforest such areas, trees must often be planted and nurtured, for which the landowner can count on spending nearly $1,000 per acre. Moutinho notes that people often forget to factor in the expense of taking care of seedlings after they're planted, which is around 60 percent of the total tab. In some places, the cost can go as high as $5,000 per acre.

For landowners, regrowing the forest above and beyond the mandates of Brazilian law would be an act of volunteerism. There could be some hardwood timber to harvest in thirty to forty years, but, Moutinho says, no one plants native trees with an eye to logging revenue. If wood is what they want from FLR, landowners opt, as Simon Lewis and company pointed out, for exotics like teak, eucalyptus, and pine in rows.

It would help if landowners could get paid for the carbon accumulating in the returning forest, but biomass is slow to come back. "I've never seen a forest grow back from degraded pasture to the way it was before. They transform from a primary forest with a canopy 40 meters (130 feet) high to something with trees that don't get taller than 6 meters (20 feet)," said Moutinho. That's because deforestation has made the *local* climate hotter and drier, and seed dispersers like tapirs have fled deep into uncut forests. What grows is a tangle of vines, skinny saplings, and grass. And then it all burns.

Blazes have been increasing in intensity and frequency for years. Set by farmers and stoked by climate change, fires are turbocharged by the microclimatic drying caused by local tree losses. In drought years since 2000, fires affected 5 to 12 percent of the southeastern Amazon, compared with less than 1 percent in a normal year. "Normal" years are becoming exceptional. A drought of a magnitude expected every hundred years hit the region in 2005, only to be followed by extreme dry weather again in 2010 and 2015. This pattern makes high-cost reforestation an even sketchier proposition. The risk of incineration can be reduced by a firebreak, a fringe of bare earth around the forest. This protection works some of the time and costs around $30 to $100 per mile annually, according to Moutinho.

Farmers have seen more encouraging results when restoring trees along the banks of the countless rivers and creeks that vein the forest landscape. Water reappears swiftly in streambeds dried out by previous deforestation, which cools the local climate and brings back large mammals that eat fruits and disperse seeds, which grow into more trees that supply the Amazon's flying rivers with their moisture. Fast-growing trees, some exotic, are mixed in for fruits, fuel, or wood. This approach mirrors one of Brazil's most epic reforestation successes. The Tijuca Forest in Rio de Janeiro is the largest urban forest in the world at around 80,000 acres. It was replanted in the nineteenth century to control erosion and protect water supplies for the city. Exotic species were mixed with natives, producing a riff on the Atlantic Forest, not a replica. The new jungle provides incalculable benefits to millions of *Cariocas* and viable habitat for some native fauna.

For degraded areas outside of river zones, Moutinho has another idea. "Let's restore the deforested areas for production." He cites World Bank estimates indicating that upping the efficiency of cattle pasture by 50 percent from its currently abysmal one head per 2.5 acres would free up 100 million acres of already deforested Amazonia for agriculture. "Instead of creating a huge program trying to reforest millions and millions of hectares, let's create a big program to rehabilitate degraded pasture to grow food so we can stop deforesting." In other words, anchor intensified agriculture to degraded zones so that it ceases to storm through the basin's primary forests.

Moutinho's pressure relief prescription fits the Bonn Challenge FLR idea of saving forests indirectly. To work, it has to be accompanied by direct protection measures. Otherwise, Amazon farmers would till restored pasture *and* deforest more land, as long as there's enough global demand for beef and soybeans and enough people available to cut down trees. Indeed, high-tech intensive farming in the United States has not prevented 99 out of every 100 acres of tallgrass prairie that once covered 170 million acres from being plowed under. What a large additional supply of Amazon farmland could accomplish is to temporarily reduce the economic pressure for deforestation. The pause could provide a window of opportunity for governments to formally protect the ample areas of undesignated intact forests.

In the Taiga, there's a similar opportunity. Russia has 190 million acres of boreal territory where forest can regrow itself, according to a study led by Ilona Zhuravleva, head of Greenpeace's Global Mapping Hub in Moscow. Her team examined the last twenty years of satellite data, eyeing all the land officially zoned for agriculture. They found an astonishing amount where forest is growing back on its own because farming isn't profitable. All people have to do is watch, and carpets of birch and pine issue forth. In the Tunka Valley near Lake Baikal, a lime-green ribbon of young trees cloaks the lower slopes, meeting taller forest at a boundary that looks like it was drawn with a ruler. It's an abandoned field. In Soviet times the government subsidized and mandated the cultivation of Siberian hills that just want to grow trees. After the collapse, trees resumed growing.

The hitch is that the reemerging trees are illegal. They may get a pass in Tunkinsky because it's a national park, but Russian law requires that most landholders keep farming such areas or face steep fines. So, when trees sprout, people burn them. The arson sends wildfires into other nearby forests and, Greenpeace asserts, represents one of the main causes of the megafires that raged across the Taiga in the summer of 2019. Allowing 190 million acres of field to become forest would sequester 350 million tons of carbon, an amount equal to France's 2017 emissions.

But Greenpeace-Russia head Alexey Yaroshenko has another, arguably more pragmatic, notion. As unlikely as this might sound coming from a lifelong campaigner for wild forests, Yaroshenko wants to promote tree farms. "The only way to save some biodiversity is to polarize logging. Leave some areas wild and use some areas in an intensive way." He estimates that by putting just two-thirds of the miscategorized "agricultural" land into tree plantations, Russia could handily meet its wood needs. While less carbon would be stored in the wood-producing areas than if they were left to fully revert to mature forests, carbon would be retained elsewhere, in the wild Russian woods from which loggers could be diverted. On balance, he says, Russia would come out ahead in terms of biodiversity, carbon, and community.

"People need jobs. The government tries to do rural development, but they use the logic of the beginning of the twentieth century. . . . Modern industrial agriculture doesn't give enough jobs because it's heavily mechanized. People, they like to live together. They want a big settlement, in order to have medicine, to have schools, to have transport, to have internet, and so on." Wood processing could support enough jobs for those sorts of towns, Yaroshenko says. He reminds us that, as in the Amazon, squeezing production from managed forests won't save natural ones automatically. "Without expanding protected areas there will not be enough motivation for loggers to move from wilderness to developed areas. Good forestry costs more than the wood mining," he says, referring to the practice of extracting timber and moving on with little regard for the logged ecosystems' capacity to recover and provide future harvests.

In one broad valley to the east of Lake Baikal, the conifers stand in a solid polygon with straight edges. A local friend, Alexey Khamaganov, confirmed as we drove past that it was a reforestation site. It reminded him of field trips he took as a kid. The schoolchildren, unleashed in a swarm, would dash around, push each other, fall, yelp, laugh, and, almost incidentally, plant trees. Forestry authorities began to notice that the trees planted by the elated youth survived in greater numbers than did the ones planted by gloomy adults. "So now the forestry vans go to schools and they fill up with children and go out to do the reforestation," Khamaganov said with a laugh.

"If you are attending this conference, you need to commit to planting 1 trillion trees."

That was the challenge issued by Salesforce CEO Marc Benioff to the tycoons, heads of state, and luminaries gathered at the 2020 World Economic Forum in Davos, Switzerland. Benioff announced at the forum that 300 companies were already on board, joining Salesforce, which vowed to plant 100 million of the trees. The Davos website fizzed with solutions, including a laborsaving drone that strafes the bare ground with germinated tree seeds and an app that allows people to fund reforestation on their smartphones. A trillion new trees would occupy about a billion acres of land. An inventory of global tree-growing real estate concluded that there's enough space.

It's a startlingly big goal. For perspective, Salesforce's impressive 100-million-tree contribution would account for just one out of every 10,000 of the seedlings. In the five years leading up to 2015, the globe was gaining about 4.3 billion trees a year and losing 7.6 billion, a trend that roughly continues to the present. If environmental efforts could stop the 7.6 billion tree losses and keep up the gains, it would take 232 years to add a trillion trees. Benioff proposed to do it in ten.

Davos goer Greta Thunberg, ever the voice of youthful straight talk, called the Trillion Trees bandwagon "a distraction." It is. Billions of dollars and political capital could be put to better use on forest-saving and other climate initiatives that are cheaper, less risky, and more beneficial to the webs of nonhuman life that sustain us. Canny entrepreneurs should readily recognize the cost effectiveness and multiple benefits of keeping carbon in the magnificently complex forests we have now, secured in parks and forest peoples' collective territories. Next on the keen forest investor's list should be protecting the places that have been lightly logged that can quickly recover as fully functioning ecosystems. And for purposes of consistency and not canceling out the benefits of their forest contributions, business titans might elect not to serve clients in the industries making things worse. Oil, coal, and irresponsible agribusiness firms could, for example, use some system other than Salesforce to keep track of their customers.

Once all that's done, we also need more forest. A lot of carbon now wandering in the atmosphere requires a home down here, which can also nurture woodland fauna and bring rain. We see a thread of true love for the planet under the bravado and alliteration of the Trillion Trees. And while most of these trees need not be individually planted, there's a place for planting in the forest recovery solution set. Planting trees is a spiritual act, a life-giving, kinship-affirming prayer-in-action, one that puts people immediately and tangibly in the role of caretaker, parent, and protector. Even those who live in a city can go out and write the first few lines of an ecological epic that may take a million twists and turns involving insects, birds, mushrooms, and wind over multiple human lifetimes. As long as we do the rest of what's needed for the megaforests, the trifling near-term impact planting has on the balance of gases in the atmosphere matters less than changes that may take place in the heart of the planter.

12

An Invitation

Tamasaimpa, trapped in a Zoom square like many of us in mid-2020, reflected on humanity's choice. "I was reading *Genesis* to my dad one time, translating the Bible into Marubo. I was reading about paradise, where Eve ate something, whatever it was, and got in trouble. And my father said, 'That's it. That's the place *Canavoã* gave us.'" Tamasaimpa paused for a moment. Canavoã is the Marubos' supreme spiritual being. "I'm telling you, there are places where paradise still exists. Sitting next to a creek hearing all the birds, that's my paradise. I'm not sure how to put it in words, but this is what I want to explain to you." Paradise, in Marubo, is *yovevõ may.* It means a sacred place of peace, stillness, and connection.

He continued: "People need to understand that the Earth, we are on it. There are real physical issues at play. . . . If we go back to the way we were before the pandemic, the end of the world is here. It all depends on us. You get it? We can create heaven or hell here. The forest is part of that decision—heaven or hell? It's the last of what Canavoã gave us." Losing our planet's intact megaforest will tip us into a climate trajectory incompatible with stable human society, extinguish millions of life-forms, and

Marubo homeland on the upper Curuçá River. *(©Sebastião Salgado)*

render humanity an ever-more uniform army of bipeds carousing on an increasingly lifeless rock. It's Tamasaimpa's hell.

The Druzhina movement of Russian student-environmentalists was launched in the early 1960s with the motto "For the success of a hopeless cause!" Anyone with a career in conservation who is even mildly introspective can relate. One wonders, from time to time, if our society isn't fated to supernova and carry a lot of our blameless nonhuman Earthlings with us. Markets, the potent invisible hand, offer automatic rewards for disassembled forest parts—timber, game, peat, and soils. The forests themselves, meanwhile, require resolute and steadfast collective action, which is challenging in a world with tempestuous politics. Even within each of us, as individuals, distractions and self-indulgence sometimes crowd out our more generous and enlightened instincts.

The Druzhina were student nature lovers trying to protect ecosystems under an authoritarian regime that could (and did) cancel dozens of parks in a day. But they embraced their "hopeless" cause with gusto and persistence and today have a lot to show for it. Similarly, people around the world continue to get up every day to nudge, cajole, and lead societies toward a promising environmental trajectory as best we can. It's important to view this as a very long project of learning, scientific and spiritual, and to consider environmental goals for a few hundred years from now. There's enough space and biological resilience and people focused on nature that we can realistically envision a distant descendant walking in a forest bigger than she can wrap her mind around, full of birds, ferns, mushrooms, fish, big cats, orchids, marsupials, and other forms of life that made it through the funnel of the climate crisis and its associated trials. In her world, we *Homo sapiens* will have grown into our power and learned to cooperate with each other. We will have pulled back from breakneck population growth and retired quaint terms like "natural resource" in favor of other ways of talking about a planet to which we're tied by a kinship so obvious that no one needs to insist on it anymore.

The patience we need for this work is exemplified by the most forest-like big building in the world, Antoni Gaudí's Sagrada Familia basilica in Barcelona. Columns dappled with multicolored sunlight branch and spread into a canopy of serrated blossoms. The architect's Catalan modernist style transcribed nature into stone, iron, and glass. Nothing like it had ever been built, and when masons laid the cornerstone in 1882, they likely had no illusions of seeing the end result. Gaudí himself died in 1926. Sagrada Familia, expected to be finished during the current decade, reminds us that such works are made by one generation after the other, each adding something to what the last one left.

Our cathedral is to learn—or relearn—how to live on Earth.

The parks, Indigenous lands, roadlessness, and forest recovery we've advocated here are some of the stones we need to add to that cathedral right away. They are the contributions of our time. If the wooded nations do succeed in passing along to the next generation an Earth with megaforests, with vast expanses of Marubo *yovevõ may*, it will be because millions of people organized and called forth the twin forces of restraint and cooperation on an unprecedented scale. A shared land ethic will have prevailed over the small plans and ephemeral indulgences of our factions.

Much of what we've proposed here is the business of bureaucrats and public officials. For the vast majority of us who don't handle the levers of forest policy, how can we live daily lives that best support the megaforests? Some contend that individual consumer choices are too small to make a difference—the change needs to start with the corporations that are digging and scraping the remotest parts of the globe. An alternative view is that corporations simply give people what they want; if people want green energy and small cars, that's what companies will provide. In this view, corporations are empty vessels, the consumer's mindless servants. Neither view is right. People who buy and people who sell are part of one system. Generalized anti-corporate finger-pointing won't help, nor will corporations' disingenuous moral neutrality. Both sides need to own this challenge.

Forestlands are fully enmeshed in our food, fiber, and energy economies in two ways: first as potential sources of raw material and places to plant. The grander our societies' material demands, the more we ask of forests. The less people use the better. Second, forests are impacted by the warming our consumer choices provoke. Fast change will certainly unravel some ecological webs and wipe out species, which evolution will take a long, long time to replace.

So, yes, our individual choices matter. Next time you build or buy a house, consider a smaller one. Less lumber means less logging in the southern boreal forest. Less copper wire means fewer mines in the mountains of New Guinea and Alaska. Fewer plastic pipes mean fewer oil wells in the Andean Amazon foothills. And so on, for everything your house is made of. You can choose a smaller car, ride a bike, decorate your roof with solar panels, turn the heat down, lights off, AC up/off, use a clothesline, fly less and video-talk more, eat less beef or none at all. Buy from sustainable sources, but don't buy more just because it's green.

In the course of writing this book, we visited various forest people and asked them what they wanted readers to know about their forests. The premise of the question was that we could take messages back and inform people who might wish to help protect the wilds of New Guinea or the distant Taiga, even though they might never visit. We thought our hosts might reveal compelling esoterica of forest remedies, spirits, and foods. Or they might relate trends of climate change drying out streams or causing land to heave with the loss of its permafrost. They might, we thought, have some instructions.

The most common message for our readers was none of these. It was an invitation: "Tell them to come!"—usually exclaimed with a big smile. So, we faithfully deliver the message: Go see a big forest! The people who live there want you to experience, directly and with all your senses, what we've done our best to hint at between these covers. Some big forests are easier to reach than you might think. So, go if you can.

It may be the case, however, that time, money, your own personal carbon budget, pandemic, or some other factor may prevent your getting into the middle of a megaforest. In that case, and even if you do have the chance to go to the Amazon, Congo, or Siberia, we have an additional suggestion.

Go outside. Frequently. Step outside anywhere and find a leaf and permit it to blow your mind. Check out its delta of veins. Run your finger on its underside. Taste it. Check if it has hair. Crumple it and smell it. Go further, to a forest of any size, a forest clearing, a clump of trees, or even a spot under a single specimen—someplace where, even though you may hear cars and dogs in the distance, you can sit on soft, uneven ground, unseen. Consider the unspooling ribbon of human affairs that the surrounding trees have witnessed and with what interest or indifference they may have watched. Inspect the ground and picture the interlaced fingers of mycelium and roots that swap sugar and water and carbon and data, a mushroom-assisted conversation that betrays care among trees. Notice the mosaic of leaves catching light or the weave of needles on the ground. Be still and birds will invade your copse. Trees, even in small groups, exhale monoterpenes that reduce stress, lower blood pressure and heart rate, and perhaps even trigger dopamine. So stay long enough to feel your mood change, watch shadows shorten or stretch. Get caught by rain or snow or nightfall. Get a little lost.

A small forest is like a fractal knob of a big one. It may not have elephants or wolves or jaguars, uncontacted humans, fossil mammoths, the world's record for biological diversity, or a 10-foot peat cake underneath. But the woods at hand are a practical place to meet the rest of creation. This meeting is important for more than mere enchantment. That's because we people are vulnerable to a delusion, namely, that we know enough about something as soon as we stop experiencing it. At a distance, real forests congeal into the static idea "forest." Contact forces observation and inquiry, rewiring us to the astonishing reality of the things that still exist.

We ride face-to-face on benches in the back of a Toyota Land Cruiser, bushes brushing its sides on a dirt road in northern Congo, destined for the trailhead to a gorilla research camp called Mondika. Riding with us is a tall young man named Thierry Fabrice Ebombi, who goes by Fabrice. He grew up and went to college in Brazzaville, hundreds of miles from the forest. In 2012, he answered an ad for a job with Dave Morgan's project studying primates in Nouabalé-Ndoki National Park. Shortly after arriving, and with no previous experience, he was unexpectedly detailed to habituate a group of gorillas. Ebombi's predecessor had quit suddenly after three years making little headway convincing the gorilla family led by the silverback Loya to let biologists hang around.

Ebombi read up on the subject before going out to face Loya. The standard approach can take a year and involves having two or more people follow the silverback male until the animal becomes so fed up that he charges the humans. When the 500-pound freight train of ape fury bears down on you, the correct technique is to stand your ground. They tend

Fabrice Ebombi at the Mondika research camp in the Republic of the Congo.

to pull up at the last minute. After a large number of repetitions, most silverbacks and their families will conclude that the humans are neither a threat nor avoidable and will tolerate, perhaps even accept the people. Or even care for them; one day a gorilla near Mondika alerted researchers to the presence of a deadly butterfly viper.

"Fabrice here is our gorilla whisperer" Dave Morgan drawls from his place on the bench next to Ebombi, who is heading out to the Mondika camp to begin habituating a new group. Morgan reveals that Ebombi has devised a new method that shortens the process to just a few months. John wonders how one might put a gorilla more quickly at ease. He imagines approaching at the correct hour, maybe with fewer people, perhaps at a certain angle, or when the family has dispersed to feed and the silverback is alone. He asks the gorilla whisperer of the Goualougo Triangle to describe the technique.

Fabrice Ebombi smiles.

"Il faut les aimer."

Which is French for "you have to love them." Ebombi doesn't elaborate. Technique matters, and no doubt the man has plenty. But his questioner is never going to attempt to habituate a silverback. There are other contributions, however, that we all might undertake, each applying our particular skills and resources, to ensure a world with life-giving megaforests. And for any such effort, Fabrice Ebombi's four words are a good place to start.

ACKNOWLEDGMENTS

Carol Andrews read and read, so many drafts that we lost count. With scissors, tape, colorful markers, X-ray vision for double meanings, and a formidable love of the forest, she steered us along a path of integrity and care much more rarified than any we could have found without her.

Bob Hepner started reading our chapters in the last year of his life. He gathered the energy to get on the phone and talk about each one, offering funny comments, earnest encouragement, and helping us omit material (including mathematical equations!) for the benefit of later readers. After walking alongside us through the first draft, Bob took his leave and made his way up into a canyon in the Organ Mountains.

We are deeply grateful to W. W. Norton and our brilliant editors there, John Glusman and Helen Thomaides. Mr. Glusman immediately saw the need for this book and assured us we could write it. The two knew just how deep to wade in and help us get it done.

We are humbled by Sebastião Salgado's contribution of several images that appear herein. For decades his photographs have made us feel places, people, and their stories in our marrow; his participation in this project is a deep honor. We also thank another outstanding Brazilian photographer,

Marcos Amend, for his arapaima shot. John's craft behind the lens was formed in large part thanks to his many opportunities to set up his tripod next to Marcos's and learn. On the book's visuals, we also recognize David Atkinson, whose sublime maps make us want to go places.

Thanks to those present at creation: to Kris Tompkins for the campsite in the Patagonia Park in Chile where some early words were written; to the late Andrew Tilin for coaching; to Amy Rosenthal for matchmaking; to John Guida and Chris Conway for lending the *New York Times*' megaphone; and to our agent Lauren Sharp, of Aevitas, for seeing a book before anyone else did.

Our hosts in the grandest of Earth's forests won our affection and deepest respect. Dave Morgan and the good people at the Wildlife Conservation Society, the Nouabalé-Ndoki National Park, and the Goualougo Triangle Ape Project shared the big beautiful intact forests of northern Congo, explained their history and nuance, and introduced us to the extraordinary nonhuman primates of the region. Many thanks to Steve Ross and Jillian Braun, of the Lincoln Park Zoo, for making a spot in their truck and pirogue for John.

Special thanks to Don Reid and Norman Barichello, who greeted John in the Yukon. Don and his colleagues Hilary Cooke and Jamie Kenyon shared science and their infectious dedication to the far north. Norman introduced us to the community of Ross River, where elders Clifford McLeod and John Acklack provided a kind welcome to Kaska Dena territory. Norman also introduced us to his son Josh, who translated from Kaska and offered insights about how language shapes not just what you can think but what you can do. Thanks to Caitlynn Beckett for sharing her research trip, and to John Ward for sharing a meal and stories in Taku River Tlingit country.

The fine team from the Samdhana Institute showed John the mountain forests and coastal jungles of West Papua in February 2020, at the last moment such things were possible. Yunus Yumte, Sandika Ariansyah, Betwel Yewam, and Agustinus Tafuran made up the crew traveling across Tambrauw. We are deeply grateful to the village of Ayapokiar and the Momo clan in particular for the chance to walk, sleep, and swim on their ancestral lands. In Kwoor, Derek Mambrasar and family offered shelter in the storm and time in their lowland forest. We also would like to recognize Gabriel Asem, Bupati of Tambrauw, for his gracious reception in Fef and for his extraordinary vision for conserving the homeland he governs.

Many thanks to Sergei Natvevich, Norbu Lama, Erjen Khamaganova, Maria Azhunova, Alexey Khamaganov, and Rinchin Garmaev for their kind welcome in Buryat and Soyot lands and for the introduction to the watery being at the center of their world, Baikal.

John sends special thanks to the peoples of the Javari Valley Indigenous Territory: the Kanamary, Korubo, Kulina, Marubo, Matsés (Mayoruna), Matís, and Tsohom Dyapá for the opportunity to work and learn alongside them in John's day job at Nia Tero. Thanks also to the dedicated people at FUNAI in the Javari.

John also sends a big and wholehearted shout out to his employer, Nia Tero, which provided time off, patience, moral support, and—most of all—an education. Thanks to you all! John is especially grateful to Chris Filardi for conversation, birdsong, and for all the intellectual and spiritual breadcrumbs. Thanks also go to John's former colleagues of many years at Conservation Strategy Fund for the superlative work and all that you taught him.

Tom extends eternal thanks to the long parade of scientists, students, and others who have shared the excitement and joy of the Amazon with us. Deep thanks to Elisabeth Cousens and the United Nations Foundation for understanding the importance of forests and to Carmen Thorndike for ever-cheerful and amazing assistance.

Several folks deserve special mention for their forbearance as we went back again and again with questions. They answered, sent us papers, connected us to others who had the answers, and read sections of the book to make sure we got things right. This group includes: Peter Potapov, Svetlana Turubanova, Valérie Courtois, Dave Morgan, Steve Kallick, Mubariq Ahmad, Tom Evans, Tamasaimpa, Bruce Beehlher, David Gordon, Dominique Bikaba, Ilona Zhuravleva, Jess Housty, Juan Chang, Jonah Busch, Paulo Moutinho, Rita Mesquita, Mario Cohn-Haft, Marcos Amend, Nonette Royo, Nadine Laporte, Mads Halfdan Lie, Vladimir Krever,

Andrey Shegolev, Michael Coe, Alexandra Aikhenvald, Victoria Tauli-Corpuz, and Jennifer Corpuz. We add our thanks to many others who provided interviews and information.

The Wildlife Conservation Society is a global standard-bearer for the cause of intact forests and was a big help in most of the regions we wrote about. Greenpeace-Russia was a terrific help with maps and studies, as

River cobble, Enatahua River, Bolivia.

was the Global Land Analysis and Discovery lab at the University of Maryland, College Park.

Roger Kent, Ronnie Karish, Nick Sanders, Darryl Berlin, and Catherine Filardi all pitched in to get the first draft of the book into shape. We had excellent research assistance early in the game from Elizabeth Hiroyasu, Virginia Ryan, and Jessica Reid. Jessica also did all the GIS work to prepare data layers, including forest base maps, for the six maps in the book. Crucially, she also read some of the more unruly chapters and, at the right times, said, "It's good, Dad. Really."

Finally, we would like to extend our humble and deepest gratitude to the people—tribes, scientists, activists, bureaucrats, and leaders—who are taking care of the woods that take care of us.

NOTES

Prologue: Anastasia's Woods

2 **Their boundaries are defined:** J. P. Brandt, "The Extent of the North American Boreal Zone," *Environmental Reviews* 17 (June 2009): 101–61, https://doi.org/10.1139/A09-004.

5 **involve reversing deforestation by 2030:** Joeri Rogelj et al., "Mitigation Pathways Compatible with 1.5°C in the Context of Sustainable Development," in *Global Warming of 1.5°C. An IPCC Special Report on the Impacts of Global Warming of 1.5°C above Pre-Industrial Levels and Related Global Greenhouse Gas Emission Pathways, in the Context of Strengthening the Global Response to the Threat of Climate Change, Sustainable Development, and Efforts to Eradicate Poverty* (Geneva: Intergovernmental Panel on Climate Change, 2018), 82.

5 **a cluster of volcanoes in Siberia:** Daniel H. Rothman et al., "Methanogenic Burst in the End-Permian Carbon Cycle," *Proceedings of the National Academy of Sciences of the USA* 111, no. 15 (April 15, 2014): 5462–67, https://doi.org/10.1073/pnas.1318106111; Hana Jurikova et al., "Permian–Triassic Mass Extinction Pulses Driven by Major Marine Carbon Cycle Perturbations," *Nature Geoscience* 13, no. 11 (November 2020): 745–50, https://doi.org/10.1038/s41561-020-00646-4; Kunio Kaiho et al., "Pulsed Volcanic Combustion Events Coincident with the End-Permian Terrestrial Disturbance and the Following Global Crisis," *Geology*, November 4, 2020, https://doi.org/10.1130/G48022.1.

5 **downsizing their genomes:** Kevin A. Simonin and Adam B. Roddy, "Genome Downsizing, Physiological Novelty, and the Global Dominance of Flowering Plants," *PLOS Biology* 16, no. 1 (January 11, 2018): e2003706, https://doi.org/10.1371/journal.pbio.2003706.

6 **unfragmented forests hold double the average carbon:** Peter Potapov et al., "The Last Frontiers of Wilderness: Tracking Loss of Intact Forest Landscapes from 2000 to 2013," *Science Advances* 3, no. 1 (January 2017): e1600821, https://doi.org/10.1126/sciadv.1600821.

6 **1.8 trillion metric tons of:** Corey J.A. Bradshaw and Ian G. Warkentin, "Global Estimates of Boreal Forest Carbon Stocks and Flux," *Global and Planetary Change* 128 (May 2015): 24–30, https://doi.org/10.1016/j.gloplacha.2015.02.004; "Global CO2 Emissions in 2019—Analysis," IEA, accessed November 16, 2020, https://www.iea.org/articles/global-co2-emissions-in-2019.

6 **carbon in tropical forests costs a fifth as much:** Jonah Busch, "What Does the Resurgence in Carbon Pricing Mean for Tropical Deforestation?," Center For Global Development, 2016, https://www.cgdev.org/blog/what-does-resurgence-carbon-pricing-mean-tropical-deforestation; Jonah Busch and Jens Engelmann, "The Future of Forests: Emissions from Tropical Deforestation with and without a Carbon Price, 2016–2050," *SSRN Electronic Journal*, 2015, https://doi.org/10.2139/ssrn.2671559.

6 **unmentioned in most national climate plans:** James E. M. Watson et al., "The Exceptional Value of Intact Forest Ecosystems," *Nature Ecology & Evolution* 2, no. 4 (April 2018): 599–610, https://doi.org/10.1038/s41559-018-0490-x; Brendan Mackey et al., "Policy Options for the World's Primary Forests in Multilateral Environmental Agreements," *Conservation Letters* 8, no. 2 (2015): 139–47, https://doi.org/10.1111/conl.12120.

6 **twenty new monkeys:** "'New Species of Monkey' Discovered in Amazon Rainforest," *The Independent*, December 23, 2019, https://www.independent.co.uk/environment/nature/monkeys-new-species-amazon-rainforest-brazil-parecis-plateau-deforestation-a9257921.html.

7 **variety of Amazonian grammars:** Robert M. W. Dixon and Alexandra Y. Aikhenvald, eds., *The Amazonian Languages*, Cambridge Language Surveys (Cambridge, UK; New York: Cambridge University Press, 1999), 1.

8 **control around a third of intact forests:** Julia E. Fa et al., "Importance of Indigenous Peoples' Lands for the Conservation of Intact Forest Landscapes," *Frontiers in Ecology and the Environment* 18, no. 3 (April 2020): 135–40, https://doi.org/10.1002/fee.2148.

9 **anywhere else in the Amazon:** Wayne S. Walker et al., "The Role of Forest Conversion, Degradation, and Disturbance in the Carbon Dynamics of Amazon Indigenous Territories and Protected Areas," *Proceedings of the National Academy of Sciences of the USA* 117, no. 6 (February 2020): 3015–3025, https://doi.org/10.1073/pnas.1913321117; Christoph Nolte et al., "Governance Regime and Location Influence Avoided Deforestation Success of Protected Areas in the Brazilian Amazon," *Proceedings of the National Academy of Sciences of the USA* 110, no. 13 (March 26, 2013): 4956–61, https://doi.org/10.1073/pnas.1214786110.

10 **called the Gila:** Philip Connors, *Fire Season: Field Notes from a Wilderness Lookout* (New York: Ecco, 2011).

Chapter 1: The Forest System

14 **"smoker's teeth":** Matt Richtel, "Now It's Not Safe at Home Either. Wildfires Bring Ashen Air into the House," *New York Times*, September 12, 2020, sec. Health, https://www.nytimes.com/2020/09/12/health/fires-air-california.html.

16 **A pair of reports:** Rogelj et al., "Mitigation Pathways Compatible with 1.5°C in the Con-

text of Sustainable Development"; *Climate Change and Land.* An IPCC special report, accessed November 30, 2019, https://www.ipcc.ch/report/srccl/.

16 **tropical forests reabsorbed around half of it:** Frances Seymour and Jonah Busch, *Why Forests? Why Now? The Science, Economics, and Politics of Tropical Forests and Climate Change* (Washington, DC: Center for Global Development, 2016).

16 **carbon budget for 1.5°C:** Rogelj et al., "Mitigation Pathways Compatible with 1.5°C in the Context of Sustainable Development."

16 **16 to 19 percent:** Seymour and Busch, *Why Forests?*

17 **tucking into telephone poles:** Cheryl Katz, "Small Pests, Big Problems: The Global Spread of Bark Beetles," *Yale Environment 360*, September 21, 2017, accessed January 3, 2021, https://e360.yale.edu/features/small-pests-big-problems-the-global-spread-of-bark-beetles.

17 **"moderate" impacts to "high" ones:** *Global Warming of 1.5 °C*, IPCC special report, accessed August 25, 2020, https://www.ipcc.ch/sr15/; *Graphics—Global Warming of 1.5 °C*, IPCC, accessed August 25, 2020, https://www.ipcc.ch/sr15/graphics/.

17 **six Asian mainland countries:** Scott A. Kulp and Benjamin H. Strauss, "New Elevation Data Triple Estimates of Global Vulnerability to Sea-Level Rise and Coastal Flooding," *Nature Communications* 10, no. 1 (October 29, 2019): 4844, https://doi.org/10.1038/s41467-019-12808-z.

17 **dialed the temperature goal down to 1.5°C:** UNFCCC, *Paris Agreement*, accessed January 6, 2021, https://unfccc.int/sites/default/files/english_paris_agreement.pdf.

17 **store 40 percent of the aboveground:** Potapov et al., "The Last Frontiers of Wilderness."

17 **New research led by Sean Maxwell:** Sean L. Maxwell et al., "Degradation and Forgone Removals Increase the Carbon Impact of Intact Forest Loss by 626%," *Science Advances* 5, no. 10 (October 1, 2019): eaax2546, https://doi.org/10.1126/sciadv.aax2546.

19 **47 to 83 percent of carbon:** Bradshaw and Warkentin, "Global Estimates of Boreal Forest Carbon Stocks and Flux"; "Peatland Fires and Carbon Emissions." Natural Resources Canada, November 8, 2012, https://www.nrcan.gc.ca/climate-change/impacts-adaptations/climate-change-impacts-forests/forest-carbon/peatland-fires-carbon-emissions/13103.

19 **154 and 197 metric tons per acre:** Bradshaw and Warkentin, "Global Estimates of Boreal Forest Carbon Stocks and Flux."

19 **a more thorough 2011 inventory:** Y. Pan et al., "A Large and Persistent Carbon Sink in the World's Forests," *Science* 333, no. 6045 (August 19, 2011): 988–93, https://doi.org/10.1126/science.1201609; Bradshaw and Warkentin, "Global Estimates of Boreal Forest Carbon Stocks and Flux."

20 **4.25 and 7.5 feet deep:** L. L. Bourgeau-Chavez et al., "Mapping Peatlands in Boreal and Tropical Ecoregions," in *Comprehensive Remote Sensing*, Vol. 6 (Amsterdam: Elsevier, 2018), 24–44, https://doi.org/10.1016/B978-0-12-409548-9.10544-5.

20 **found 124 new species:** James Gorman, "Is This the World's Most Diverse National Park?," *New York Times*, May 22, 2018, sec. Science, https://www.nytimes.com/2018/05/22/science/bolivia-madidi-national-park.html; "Two-and-a-Half-Year Expedition Ends in World's Most Biodiverse Protected Area: Identidad Madidi Finds 124 Taxa That Are

Candidates for New Species to Science," ScienceDaily, accessed December 23, 2020, https://www.sciencedaily.com/releases/2018/05/180522132607.htm.

21 **small number of species overall:** Charles Darwin, *On the Origin of Species*, illustrated ed. (New York: Sterling, 2008), 391.

21 **metaphor for the fragments of habitat:** Robert H. MacArthur and Edward O. Wilson, *The Theory of Island Biogeography* (Princeton, NJ: Princeton University Press, 1967), 3–4.

21 **Barro Colorado's managers accurately declare:** Smithsonian Institution, "Barro Colorado," Smithsonian Tropical Research Institute (Smithsonian Tropical Research Institute, October 31, 2016), https://stri.si.edu/facility/barro-colorado.

22 **a simple answer about fragmentation:** William F. Laurance et al., "Ecosystem Decay of Amazonian Forest Fragments: A 22-Year Investigation," *Conservation Biology* 16, no. 3 (2002): 605–18, https://doi.org/10.1046/j.1523-1739.2002.01025.x.

22 **three times too small for the birds:** Laurance et al., "Ecosystem Decay of Amazonian Forest Fragments."

24 **couple hundred feet of treeless ground:** Laurance et al., "Ecosystem Decay of Amazonian Forest Fragments."

24 **the park had over 3,000:** Terry Brncic, "Results of the 2016–2017 Large Mammal Survey of the Ndoki-Likouala Landscape" (Wildlife Conservation Society, n.d.).

25 **15 percent more carbon:** Fabio Berzaghi et al., "Carbon Stocks in Central African Forests Enhanced by Elephant Disturbance," *Nature Geoscience* 12, no. 9 (September 2019): 725–29, https://doi.org/10.1038/s41561-019-0395-6.

25 **Another study in the vicinity:** John R. Poulsen, Connie J. Clark, and Todd M. Palmer, "Ecological Erosion of an Afrotropical Forest and Potential Consequences for Tree Recruitment and Forest Biomass," *Biological Conservation*, Special Issue: Defaunation's Impact in Terrestrial Tropical Ecosystems, 163 (July 1, 2013): 122–30, https://doi.org/10.1016/j.biocon.2013.03.021.

26 **26 to 38 percent are possible:** Carlos A. Peres et al., "Dispersal Limitation Induces Long-Term Biomass Collapse in Overhunted Amazonian Forests," *Proceedings of the National Academy of Sciences of the USA* 113, no. 4 (January 26, 2016): 892–97, https://doi.org/10.1073/pnas.1516525113.

26 **heroics in data collection:** Lucas N. Paolucci et al., "Lowland Tapirs Facilitate Seed Dispersal in Degraded Amazonian Forests," *Biotropica* 51, no. 2 (2019): 245–52, https://doi.org/10.1111/btp.12627.

26 **fill the ecological gap:** Lera Miles et al., "A Safer Bet for REDD+: Review of the Evidence on the Relationship between Biodiversity and the Resilience of Forest Carbon Stocks," Working paper v2, Multiple Benefits Series (Cambridge, UK: UN-REDD Programme, UNEP World Conservation Monitoring Centre, 2010).

26 **even in the bodies of elephants:** Ralph Chami et al., "On Valuing Nature-Based Solutions to Climate Change: A Framework with Application to Elephants and Whales," *SSRN Electronic Journal*, 2020, https://doi.org/10.2139/ssrn.3686168.

26 **seeds of which are wind-borne:** Anand M. Osuri et al., "Contrasting Effects of Defaunation on Aboveground Carbon Storage across the Global Tropics," *Nature Communications* 7 (April 25, 2016): 11351, https://doi.org/10.1038/ncomms11351.

28 **juice the jungle:** Christer Jansson et al., "Phytosequestration: Carbon Biosequestration by Plants and the Prospects of Genetic Engineering," *BioScience* 60, no. 9 (October 2010): 685–96, https://doi.org/10.1525/bio.2010.60.9.6.

28 **snow reflects sunlight in winter:** G. Bala et al., "Combined Climate and Carbon-Cycle Effects of Large-Scale Deforestation," *Proceedings of the National Academy of Sciences of the USA* 104, no. 16 (April 17, 2007): 6550–55, https://doi.org/10.1073/pnas.0608998104.

28 **razing it incrementally would emit more carbon:** Deborah Lawrence et al., "Biophysical Effects of Forests on Climate: Toward a More Complete View of Climate Mitigation" (unpublished manuscript, 2020).

28 **free of surprise side effects:** S. Yachi and M. Loreau, "Biodiversity and Ecosystem Productivity in a Fluctuating Environment: The Insurance Hypothesis," *Proceedings of the National Academy of Sciences of the USA* 96, no. 4 (February 16, 1999): 1463–68, https://doi.org/10.1073/pnas.96.4.1463; Clarence L. Lehman and David Tilman, "Biodiversity, Stability, and Productivity in Competitive Communities," *American Naturalist* 156, no. 5 (2000): 19.

Chapter 2: Mapping the Root Forests

30 **identified the world's "frontier forests":** Dirk Bryant, Daniel Nielsen, and Laura Tangley, *The Last Frontier Forests: Ecosystems & Economies on the Edge: What Is the Status of the World's Remaining Large, Natural Forest Ecosystems?* (Washington, DC: World Resources Institute, Forest Frontiers Initiative, 1997), 11.

34 **a size that can accommodate natural processes:** Alexey Yu Yaroshenko, Peter V. Potapov, and Svetlana A. Turubanova, *The Last Intact Forest Landscapes of Northern European Russia* (Moscow: Greenpeace, Russia, 2001), 28.

34 **all of European Russia in 2001:** Dmitry Aksenov and Yekaterina Belozerova, *Atlas of Russia's Intact Forest Landscapes* (Moscow; Washington, DC: Global Forest Watch, 2002).

35 **NASA cut the price to $475 in 1999:** Tony Reichhardt, "Research to Benefit from Cheaper Landsat Images," *Nature* 400, no. 6746 (August 1, 1999): 702, https://doi.org/10.1038/23324.

35 **lost or fragmentated from 2000 to 2016:** Potapov et al., "The Last Frontiers of Wilderness"; Peter Potapov, "IFL_2016_national_share" (Excel worksheet provided to authors, May 20, 2020).

35 **need far more than 125,000 acres:** R. Woodroffe, "Edge Effects and the Extinction of Populations Inside Protected Areas," *Science* 280, no. 5372 (June 26, 1998): 2126–28, https://doi.org/10.1126/science.280.5372.2126.

35 **Canadian scientists note:** L.A. Venier et al., "A Review of the Intact Forest Landscape Concept in the Canadian Boreal Forest: Its History, Value, and Measurement," *Environmental Reviews* 26, no. 4 (December 2018): 369–77, https://doi.org/10.1139/er-2018-0041.

36 **size doesn't *cause* there to be more species:** MacArthur and Wilson, *The Theory of Island Biogeography*, 8.

37 **In 2020 another version of this quest:** H. S. Grantham et al., "Anthropogenic Modifica-

tion of Forests Means Only 40% of Remaining Forests Have High Ecosystem Integrity," *Nature Communications* 11, no. 1 (December 8, 2020): 5978, https://doi.org/10.1038/s41467-020-19493-3.

Chapter 3: The North Woods

45 **"untraversable forest":** "Taiga—Wiktionary," accessed August 25, 2020, https://en.wiktionary.org/wiki/taiga#cite_note-3.

45 **2 billion acres:** Kullervo Kuusela, "The Boreal Forests: An Overview," *Unasylva* 43, no. 170 (1992/3), http://www.fao.org/3/u6850e/u6850e03.htm; Taiga Rescue Network, "Boreal Forest Fact Sheet," accessed July 21, 2020, https://pdfs.semanticscholar.org/7dc1/0b6ddc8e6ac9791c9d784155cbd61b5e3dd6.pdf.

45 **The forest is mainly a mix:** S. A. Bartelev et al., *Russia's Forests: Dominating Forest Types and Their Canopy Density* (Moscow: Russian Academy of Sciences; Global Forest Watch; Greenpeace Russia, 2006), http://www.forestforum.ru/info/pictures/engmap.pdf.

46 **birches grow back first:** E.-D. Schulze et al., "Succession after Stand Replacing Disturbances by Fire, Wind Throw, and Insects in the Dark Taiga of Central Siberia," *Oecologia* 146, no. 1 (November 1, 2005): 77–88, https://doi.org/10.1007/s00442-005-0173-6.

47 **72 degrees of latitude:** Anatoly P. Abaimov et al., *Variability and Ecology of Siberian Larch Species*, technical report (Uppsala, Sweden: Swedish University of Agricultural Sciences, Department of Silviculture, 1998).

47 **toxic to wood-eating fungi:** Martti Venäläinen et al., "Decay Resistance, Extractive Content, and Water Absorption Capacity of Siberian Larch (*Larix sibirica* Lebed.) Heartwood Timber," *Holzforschung* 60, no. 1 (January 1, 2006): 99–103, https://doi.org/10.1515/HF.2006.017.

47 **oak, maple, yew, linden, ash, black birch, and poplar:** V. K. Arsenyev, *Across the Ussuri Kray: Travels in the Sikhote-Alin Mountains*, trans. Jonathan C. Slaght (Bloomington: Indiana University Press, 2016).

48 **their nest holes:** Jonathan C. Slaght, *Owls of the Eastern Ice: A Quest to Find and Save the World's Largest Owl* (New York: Farrar, Straus and Giroux, 2020), 27–8, 32–3.

48 **more than a few tight spots:** Arsenyev, *Across the Ussuri Kray*, 84, 295, 370.

48 **down 10 percent from the year 2000:** Potapov, "IFL_2016_national_share."

48 **21 percent to fire:** Potapov et al., "The Last Frontiers of Wilderness."

48 **other markers of human presence:** "Wildfires in Russia," accessed September 28, 2019, https://maps.greenpeace.org/maps/research/en.

48 **5.9 million acres:** "Wildfires in Russia."

49 **$62 fine was roughly 20 percent:** "List of Federal Subjects of Russia by GDP per Capita," in *Wikipedia*, August 15, 2020, https://en.wikipedia.org/w/index.php?title=List_of_federal_subjects_of_Russia_by_GDP_per_capita&oldid=973107377.

49 **concentrations of brown (grizzly) bears:** Josh Newell, *The Russian Far East: A Reference Guide for Conservation and Development*, 2nd ed. (McKinleyville, CA: Daniel & Daniel, 2004), 342.

50 **1.5 billion acres:** Jeff Wells, Diana Stralberg, and David Childs, *Boreal Forest Refuge:*

Conserving North America's Bird Nursery in the Face of Climate Change (Seattle, WA: Boreal Songbird Inititiative, 2018), 6.

50 **50 million acres of the United States and Canada:** *Coastal Temperate Rain Forests: Ecological Characteristics, Status and Distribution Worldwide* (Portland, OR: Ecotrust and Conservation International, 1992), accessed June 30, 2020, http://archive.ecotrust .org/publications/ctrf.html.

51 **5.8 percent loss since 2000:** Potapov et al., "The Last Frontiers of Wilderness"; Potapov, "IFL_2016_national_share."

54 **1,500 miles apart:** Matt Bowser, "Spruce Mast Events: Feast or Famine," *US Fish and Wildlife Service Kenai National Wildlife Refuge Notebook* 16, no. 31 (2014): 1–2.

54 **threadlike fungal connectors:** Suzanne W. Simard and Daniel M. Durall, "Mycorrhizal Networks: A Review of Their Extent, Function, and Importance," *Canadian Journal of Botany* 82, no. 8 (August 2004): 1140–65, https://doi.org/10.1139/b04-116.

54 **10,000 years later:** C. R. Stokes, "Deglaciation of the Laurentide Ice Sheet from the Last Glacial Maximum," *Cuadernos de Investigación Geográfica* 43, no. 2 (September 15, 2017): 377, https://doi.org/10.18172/cig.3237.

54 **75 to 100 percent of the "land":** Environment and Climate Change Canada, *Extent of Canada's Wetlands*, 2016, http://epe.lac-bac.gc.ca/100/201/301/weekly_acquisitions_list -ef/2016/16-33/publications.gc.ca/collections/collection_2016/eccc/En4-281-2016-eng .pdf.

54 **billion boreal birds winter:** Scott Weidensaul and Jeffrey V. Wells, "Saving Canada's Boreal Forest," *New York Times*, May 30, 2015, sec. Opinion. https://www.nytimes .com/2015/05/30/opinion/saving-canadas-boreal-forest.html; Wells et al., *Boreal Forest Refuge*, 6–7; "Boreal Birds," Boreal Songbird Initiative, February 28, 2014, https://www .borealbirds.org/boreal-birds.

55 **losses drop by over 70 percent:** Potapov et al., "The Last Frontiers of Wilderness"; Potapov, "IFL_2016_national_share."

58 **in the inland west:** Chelene C. Hanes et al., "Fire-Regime Changes in Canada over the Last Half Century," *Canadian Journal of Forest Research* 49, no. 3 (March 2019): 256–69, https://doi.org/10.1139/cjfr-2018-0293; Eric S. Kasischke and Merritt R. Turetsky, "Recent Changes in the Fire Regime across the North American Boreal Region—Spatial and Temporal Patterns of Burning across Canada and Alaska," *Geophysical Research Letters* 33, no. 9 (2006): L09703, https://doi.org/10.1029/2006GL025677.

58 **too young to provide seeds:** Carissa D. Brown and Jill F. Johnstone, "Once Burned, Twice Shy: Repeat Fires Reduce Seed Availability and Alter Substrate Constraints on *Picea mariana* Regeneration," *Forest Ecology and Management* 266 (February 2012): 34–41, https://doi.org/10.1016/j.foreco.2011.11.006.

59 **scald forest into a shrubbery:** Jill F. Johnstone et al., "Fire, Climate Change, and Forest Resilience in Interior Alaska," *Canadian Journal of Forest Research* 40, no. 7 (July 2010): 1302–12, https://doi.org/10.1139/X10-061.

60 **"cliomes," coined:** "Alaska-Canada Climate-Biome Shifts," SNAP, accessed December 6, 2020, https://uaf-snap.org/project/alaska-canada-climate-biome-shifts/.

60 **lose seven of eighteen cliomes:** Erika L. Rowland et al., "Examining Climate-Biome

('Cliome') Shifts for Yukon and Its Protected Areas," *Global Ecology and Conservation* 8 (October 1, 2016): 1–17, https://doi.org/10.1016/j.gecco.2016.07.006.

60 **a 2010 *PLOS ONE* paper:** Mark G. Anderson and Charles E. Ferree, "Conserving the Stage: Climate Change and the Geophysical Underpinnings of Species Diversity," ed. Justin Wright, *PLOS ONE* 5, no. 7 (July 14, 2010): e11554, https://doi.org/10.1371/journal.pone.0011554.

60 **Canada's number two toxic cleanup site:** Government of Canada, Indigenous and Northern Affairs Canada, *Remediating Faro Mine in the Yukon*, report, November 24, 2016, https://www.rcaanc-cirnac.gc.ca/eng/1480019546952/1537554989037; Dave Croft, "Massive Faro Mine Clean-up Will Begin in 2022, Two Decades after Closure," CBC, June 27, 2017, https://www.cbc.ca/news/canada/north/faro-mine-remediation-1.4179016; "Whitehorse Daily Star: Cost of Faro Project Forecast to Exceed $500 Million This Year," *Whitehorse Daily Star*, accessed March 15, 2021, https://www.whitehorsestar.com/News/cost-of-faro-project-forecast-to-exceed-500-million-this-year.

62 **$40 million per year:** Croft, "Massive Faro Mine Clean-up Will Begin in 2022."

62 **level considered safe:** Government of Canada, Indigenous and Northern Affairs Canada, *Faro Mine Remediation Project: Surface Water Quality Monitoring Baseline Report, Winter 2018*, report, August 10, 2018, https://www.rcaanc-cirnac.gc.ca/eng/1533911742197/1537556751688#chp5.

62 **58 million fish:** Alaska Department of Fish and Game, "Bristol Bay Daily Salmon Run Summary," accessed August 31, 2020, https://www.adfg.alaska.gov/index.cfm?adfg=commercialbyareabristolbay.harvestsummary.

62 **complex up to nine times bigger:** Henry Fountain, "An Alaska Mine Project Might Be Bigger Than Acknowledged," *New York Times*, September 21, 2020, sec. Climate, https://www.nytimes.com/2020/09/21/climate/pebble-mine-alaska.html.

63 **overharvested by outsiders:** Jackie Hong, "RRDC to Require Non-Kaska Hunters in Ross River Area to Get Special Permit," *Yukon News*, June 22, 2018, https://www.yukon-news.com/news/rrdc-to-require-non-kaska-hunters-in-ross-river-area-to-get-special-permit/.

63 **million miles of seismic lines:** Liam Harrap, "Fractured Forest," *Alberta Views—The Magazine for Engaged Citizens* (blog), May 1, 2020, https://albertaviews.ca/fractured-forest/.

63 **"single largest landscape disturbance":** Environment and Natural Resources, Government of the Northwest Territories, "8.2 Seismic Line Density," 2, accessed August 24, 2020, https://www.enr.gov.nt.ca/en/node/2414.

63 **Thirty-six of Canada's fifty-one boreal herds:** Environment and Climate Change Canada, *Woodland Caribou, Boreal Population (*Rangifer tarandus caribou*): Amended Recovery Strategy Proposed 2019*, November 27, 2020, https://www.canada.ca/en/environment-climate-change/services/species-risk-public-registry/recovery-strategies/woodland-caribou-boreal-2019.html#toc6.

63 **wolves and human hunters more lethal:** Environment and Climate Change Canada, *Woodland Caribou (*Rangifer tarandus caribou*): Recovery Strategy Progress Report 2012 to 2017*, program results; September 29, 2017, https://www.canada.ca/en/environment-climate-change/services/species-risk-public-registry/recovery-strategies/woodland

-caribou-report-2012-2017.html; Dave Hervieux et al., "Managing Wolves (*Canis lupus*) to Recover Threatened Woodland Caribou (*Rangifer tarandus caribou*) in Alberta," *Canadian Journal of Zoology* 92, no. 12 (December 2014): 1029–37, https://doi.org/10.1139/cjz-2014-0142; Robert Serrouya et al., "Saving Endangered Species Using Adaptive Management," *Proceedings of the National Academy of Sciences of the USA* 116, no. 13 (March 26, 2019): 6181–86, https://doi.org/10.1073/pnas.1816923116; A. David M. Latham et al., "Movement Responses by Wolves to Industrial Linear Features and Their Effect on Woodland Caribou in Northeastern Alberta," *Ecological Applications* 21, no. 8 (December 2011): 2854–65, https://doi.org/10.1890/11-0666.1.

64 **quicker to release CO_2:** Xanthe J. Walker et al., "Increasing Wildfires Threaten Historic Carbon Sink of Boreal Forest Soils," *Nature* 572, no. 7770 (August 2019): 520–23, https://doi.org/10.1038/s41586-019-1474-y.

Chapter 4: The Jungles

68 **1,028 species of birds:** Gorman, "Is This the World's Most Diverse National Park?"

68 **10 percent of Earth's species:** "Amazon | Places | WWF," World Wildlife Fund, accessed January 26, 2020, https://www.worldwildlife.org/places/amazon.

68 **are over 1,000 miles:** "Amazon | Places | WWF."

70 **organic acids leached from leaves:** Michael Goulding, Ronaldo Barthem, and Efrem Jorge Gondim Ferreira, *The Smithsonian Atlas of the Amazon* (Washington, DC: Smithsonian Books, 2003), 215–16.

72 **7 trillion gallons:** Michael T. Coe et al., "The Hydrology and Energy Balance of the Amazon Basin," in *Interactions between Biosphere, Atmosphere and Human Land Use in the Amazon Basin*, ed. Laszlo Nagy, Bruce R. Forsberg, and Paulo Artaxo, Ecological Studies (Berlin, Heidelberg: Springer, 2016), 35–53, https://doi.org/10.1007/978-3-662-49902-3_3.

72 **Eneas Salati in the 1970s:** Eneas Salati et al., "Recycling of Water in the Amazon Basin: An Isotopic Study," *Water Resources Research* 15, no. 5 (1979): 1250–58, https://doi.org/10.1029/WR015i005p01250.

72 **every country on the continent except Chile:** Thomas E. Lovejoy and Carlos Nobre, "Amazon Tipping Point," *Science Advances* 4, no. 2 (February 2018): eaat2340, https://doi.org/10.1126/sciadv.aat2340.

72 **40,000 plants:** José Maria Cardoso Da Silva, Anthony B. Rylands, and Gustavo A. B. Da Fonseca, "The Fate of the Amazonian Areas of Endemism," *Conservation Biology* 19, no. 3 (2005): 689–94, https://doi.org/10.1111/j.1523-1739.2005.00705.x; "Amazon | Places | WWF."

74 **elaborately patterned harlequin toad:** Rafael F. Jorge, Miquéias Ferrão, and Albertina P. Lima, "Out of Bound: A New Threatened Harlequin Toad (Bufonidae, *Atelopus*) from the Outer Borders of the Guiana Shield in Central Amazonia Described through Integrative Taxonomy," *Diversity* 12, no. 8 (August 2020): 310, https://doi.org/10.3390/d12080310.

76 **1910 the rubber price peaked:** Warren Dean, *Brazil and the Struggle for Rubber: A Study in Environmental History*, 1st paperback ed., Studies in Environment and History (Cambridge, UK: Cambridge University Press, 1987), 16–24, 46.

76 **1920 its bottom fell out:** Zephyr Frank and Aldo Musacchio,"The International Natural

Rubber Market, 1870–1930," EH.net, accessed February 6, 2020, https://eh.net/encyclopedia/the-international-natural-rubber-market-1870-1930/.

77 **losses are due to farms and ranches:** Potapov et al., "The Last Frontiers of Wilderness."

78 **accelerated by 34 percent:** "A Taxa Consolidada de Desmatamento Por Corte Raso Para Os Nove Estados Da Amazônia Legal (AC, AM, AP, MA, MT, PA, RO, RR e TO) Em 2019 é de 10.129 Km2" [The consolidated rate of clear-cut deforestation for the nine states of the legal Amazon (AC, AM, AP, MA, MT, PA, RO, RR and TO) in 2019 is 10,129 km2], Coordenação-Geral de Observação Da Terra, Instituto Nacional de Pesquisas Espaciais [General coordination of Earth observation, National Institute for Space Research], accessed August 23, 2020, http://www.obt.inpe.br/OBT/noticias-obt-inpe/a-taxa-consolidada-de-desmatamento-por-corte-raso-para-os-nove-estados-da-amazonia-legal-ac-am-ap-ma-mt-pa-ro-rr-e-to-em-2019-e-de-10-129-km2.

78 **a twelve-year high:** Reuters, "Brazil Amazon Deforestation Hits 12-Year High Under Bolsonaro," *New York Times*, November 30, 2020, sec. World, https://www.nytimes.com/2020/11/30/world/americas/brazil-amazon-rainforest-deforestation.html.

83 **biggest trees by volume:** Yadvinder Malhi et al., "African Rainforests: Past, Present and Future," *Philosophical Transactions of the Royal Society B: Biological Sciences* 368, no. 1625 (September 5, 2013): 20120312, https://doi.org/10.1098/rstb.2012.0312.

83 **on the adjacent plateaus:** "The Congo Basin Forest," Global Forest Atlas, accessed January 27, 2020, https://globalforestatlas.yale.edu/region/congo.

83 **Albertine Rift foothill and montane forests:** "Congo Basin Ecoregions," Global Forest Atlas, accessed January 27, 2020, https://globalforestatlas.yale.edu/congo/ecoregions/congo-basin-ecoregion.

83 **is a species treasure chest:** Andrew J. Plumptre et al., "The Biodiversity of the Albertine Rift," *Biological Conservation* 134, no. 2 (January 2007): 178–94, https://doi.org/10.1016/j.biocon.2006.08.021.

84 **173 to 251 people per square mile:** "Sud-Kivu—Democratic Republic of the Congo | Data and Statistics," Knoema, accessed May 28, 2021, https://knoema.com//atlas/Democratic-Republic-of-the-Congo/Sud-Kivu; "Nord-Kivu—Democratic Republic of the Congo | Data and Statistics," Knoema, accessed May 28, 2021, https://knoema.com//atlas/Democratic-Republic-of-the-Congo/Nord-Kivu; "Congo (Dem. Rep.): Provinces, Major Cities & Towns—Population Statistics, Maps, Charts, Weather and Web Information," accessed May 28, 2021, https://www.citypopulation.de/en/drcongo/cities/.

84 **22 humans for every square mile:** "Population Density (People per Sq. Km of Land Area)—Gabon," data, accessed July 15, 2020, https://data.worldbank.org/indicator/EN.POP.DNST?locations=GA.

84 **1.4 percent and 1 percent, respectively:** Rhett A. Butler, "The Congo Rainforest," Mongabay, accessed October 5, 2019, https://rainforests.mongabay.com/congo/.

84 **bemba pollen dating back 2,700 years:** Jefferson S. Hall et al., "Resource Acquisition Strategies Facilitate *Gilbertiodendron dewevrei* Monodominance in African Lowland Forests," ed. James Dalling, *Journal of Ecology* 108, no. 2 (March 2020): 433–48, https://doi.org/10.1111/1365-2745.13278; Carolina Tovar et al., "Tropical Monodominant Forest Resilience to Climate Change in Central Africa: A *Gilbertiodendron dewevrei* Forest Pollen

Record over the Past 2,700 Years," *Journal of Vegetation Science* 30, no. 3 (2019): 575–86, https://doi.org/10.1111/jvs.12746.

84 **30 billion tons of carbon:** Greta C. Dargie et al., "Age, Extent and Carbon Storage of the Central Congo Basin Peatland Complex," *Nature* 542 (January 11, 2017): 86.

85 **called the Congo Free State:** Adam Hochschild, *King Leopold's Ghost: A Story of Greed, Terror, and Heroism in Colonial Africa* (Boston: Houghton Mifflin, 1998).

85 **imbecile rapacity:** Joseph Conrad, *Heart of Darkness*, ed. Robert Kimbrough, 2nd ed, a Norton Critical Edition (New York: W. W. Norton, 1971), 23, 69; Hochschild, *King Leopold's Ghost*, 49.

86 **in the government's enterprises:** Hochschild, *King Leopold's Ghost.*

87 **tree can do just about anything:** Adetunji J Aladesanmi, "*Tetrapleura tetraptera*: Molluscicidal Activity and Chemical Constituents," *African Journal of Traditional, Complementary, and Alternative Medicines* 4, no. 1 (August 28, 2006): 23–36.

88 **known to use plants medicinally:** "Do Gorillas Use Plants as Medicine?," Dian Fossey Gorilla Fund, August 9, 2009, https://gorillafund.org/do-gorillas-use-plants-as-medicine/; Don Cousins and Michael A Huffman, "Medicinal Properties in the Diet of Gorillas: An Ethno-Pharmacological Evaluation," *African Study Monographs* 23, no. 2 (June 2002): 65–89; Joel Shurkin, "News Feature: Animals That Self-Medicate," *Proceedings of the National Academy of Sciences of the USA* 111, no. 49 (December 9, 2014): 17339–41, https://doi.org/10.1073/pnas.1419966111.

90 **They tickle, prank, and laugh:** Jeanna Bryner, "8 Human-Like Behaviors of Primates," livescience.com, July 29, 2011, accessed January 2, 2021, https://www.livescience.com/15309-humanlike-behaviors-primates.html.

90 **twenty-two different instances of tool use:** C Sanz and D Morgan, "Chimpanzee Tool Technology in the Goualougo Triangle, Republic of Congo," *Journal of Human Evolution* 52, no. 4 (April 2007): 420–33, https://doi.org/10.1016/j.jhevol.2006.11.001.

90 **one of these trees in every 1,500 acres:** S. T. Ndolo Ebika et al., "*Ficus* Species in the Sangha Trinational, Central Africa," *Edinburgh Journal of Botany* 75, no. 3 (November 2018): 377–420, https://doi.org/10.1017/S0960428618000173.

94 **strong associations and separate cultures:** Serge Bahuchet, "Changing Language, Remaining Pygmy," *Human Biology* 84, no. 1 (February 2012): 11–43, https://doi.org/10.3378/027.084.0101.

94 **90 percent of all timber:** Sam Lawson, *Illegal Logging in the Democratic Republic of the Congo*, Chatham House, April 2014, 2, https://www.chathamhouse.org/sites/default/files/home/chatham/public_html/sites/default/files/20140400LoggingDRCLawson.pdf.

94 **In 2018, the DRC began reinstating:** A. A. Warsame, "Democratic Republic of the Congo's Government Reinstates Illegal Logging Concessions in Breach of Its Own Moratorium," *Mareeg.Com Somalia, World News and Opinion.* (blog), February 20, 2018, https://mareeg.com/democratic-republic-of-the-congos-government-reinstates-illegal-logging-concessions-in-breach-of-its-own-moratorium/.

94 **intensifies hunting on nearly 30 percent:** N. T. Laporte et al., "Expansion of Industrial Logging in Central Africa," *Science* 316, no. 5830 (June 8, 2007): 1451, https://doi.org/10.1126/science.1141057.

94 **complicit in commercial hunting of great apes:** Dale Peterson, *Eating Apes* (Berkeley: University of California Press, 2004), 46, 156–60.

95 **Both ape species remained abundant:** David Morgan et al., "Impacts of Selective Logging and Associated Anthropogenic Disturbance on Intact Forest Landscapes and Apes of Northern Congo," *Frontiers in Forests and Global Change* 2 (July 3, 2019): 28, https://doi.org/10.3389/ffgc.2019.00028.

95 **fragmented IFLs equally or more:** Potapov et al., "The Last Frontiers of Wilderness."

95 **FSC passed a motion:** Forest Stewardship Council, General Assembly, September 2014, *Report on Results of Motions Voted on at the 2014 General Assembly*, October, 2014, 11–12.

97 **active regulation and patrols:** Gillian L. Galford et al., "Will Passive Protection Save Congo Forests?," *PLOS ONE* 10, no. 6 (June 24, 2015): e0128473, https://doi.org/10.1371/journal.pone.0128473.

97 **first European to spot it, in 1511:** Much of our description of New Guinea is based on Bruce M. Beehler, *New Guinea: Nature and Culture of Earth's Grandest Island* (Princeton, NJ: Princeton University Press, 2020), 40.

100 **westward at a similar speed:** Hugh L. Davies, "The Geology of New Guinea—The Cordilleran Margin of the Australian Continent," *Episodes* 35, no. 1 (March 1, 2012): 87–102, https://doi.org/10.18814/epiiugs/2012/v35i1/008.

100 **Geographers lump:** Beehler, *New Guinea*, 27.

101 **pieces all fell into place:** John Langdon Brooks, *Just before the Origin: Alfred Russel Wallace's Theory of Evolution* (New York: Columbia University Press, 1984), 181–83.

101 **adding to the scientific plant catalog:** Beehler, *New Guinea*, 123.

101 **elaborating new species:** Beehler, *New Guinea*,123–24.

101 **name the island:** Beehler, *New Guinea*, 18.

103 **skins became a big business:** Beehler, *New Guinea*, 49–51.

103 **a change in fashions:** William Souder, "How Two Women Ended the Deadly Feather Trade," *Smithsonian Magazine*, March 2013, accessed January 29, 2020, https://www.smithsonianmag.com/science-nature/how-two-women-ended-the-deadly-feather-trade-23187277/.

103 **degraded but not entirely cleared:** Phil L. Shearman et al., "Forest Conversion and Degradation in Papua New Guinea 1972–2002," *Biotropica* 41, no. 3 (May 2009): 379–90, https://doi.org/10.1111/j.1744-7429.2009.00495.x.

103 **fell by around 14 percent:** Jane E. Bryan and Phil L. Shearman, eds., *The State of the Forests of Papua New Guinea 2014: Measuring Change over the Period 2002–2014* (Port Moresby: University of Papua New Guinea, 2015), 17.

103 **Seventeen percent of PNG's IFL:** Potapov, "IFL_2016_national_share."

103 **road network by over 50 percent:** Mohammed Alamgir et al., "Infrastructure Expansion Challenges Sustainable Development in Papua New Guinea," ed. Govindhaswamy Umapathy, *PLOS ONE* 14, no. 7 (July 24, 2019): e0219408, https://doi.org/10.1371/journal.pone.0219408; Papau New Guinea Department of National Planning & Monitoring, "Medium Term Development Plan III (2018–2022) Volume 1," December 11, 2018, https://png-data.sprep.org/dataset/medium-term-development-plan-iii-2018-2022-volume-1; Papau New Guinea Department of National Planning & Monitoring, "Medium Term

Development Plan III (2018–2022) Volume 2," December 11, 2018, https://png-data.sprep.org/dataset/medium-term-development-plan-iii-2018-2022-volume-2.

104 **A separate 2019 analysis:** David Gaveau, "Drivers of Forest Loss in Papua and West Papua," Center for International Forestry Research, 2019.

104 **overtaking Brazil as the world leader:** Belinda Arunarwati Margono et al., "Primary Forest Cover Loss in Indonesia over 2000–2012," *Nature Climate Change* 4, no. 8 (August 2014): 730–35, https://doi.org/10.1038/nclimate2277.

104 **The mine has provided:** Susan Schulman, "The $100bn Gold Mine and the West Papuans Who Say They Are Counting the Cost," *The Guardian*, November 2, 2016, sec. Global development, https://www.theguardian.com/global-development/2016/nov/02/100-bn-dollar-gold-mine-west-papuans-say-they-are-counting-the-cost-indonesia; Jane Perlez and Raymond Bonner, "Below a Mountain of Wealth, a River of Waste" *New York Times*, December 27, 2005, sec. World, https://www.nytimes.com/2005/12/27/world/asia/below-a-mountain-of-wealth-a-river-of-waste.html.

105 **4.5 million acres of oil palm:** Gaveau, "Drivers of Forest Loss in Papua and West Papua."

105 **surprise announcement:** "Indonesia's Point Man for Palm Oil Says No More Plantations in Papua," Mongabay Environmental News, March 2, 2020, https://news.mongabay.com/2020/03/indonesia-palm-oil-papua-plantations-moratorium-luhut-pandjaitan/.

Chapter 5: Forests of Thought

113 **"an old-growth of the mind":** "Wade Davis, Why Indigenous Languages Matter," *Canadian Geographic*, October 1, 2019, accessed August 30, 2020, https://www.canadiangeographic.ca/article/why-indigenous-languages-matter.

113 **biography of his mentor:** Wade Davis, *One River: Explorations and Discoveries in the Amazon Rain Forest* (New York: Simon & Schuster, 1996), 218.

114 **43 percent of languages are endangered:** "UNESCO Atlas of the World's Languages in Danger," accessed March 9, 2021, http://www.unesco.org/languages-atlas/; Lyle Campbell et al., "New Knowledge: Findings from the *Catalogue of Endangered Languages* ('ELCat')," 3rd International Conference on Language Documentation & Conservation, Hawai'i, February 28–March 3, 2013, 17.

114 **7,139 entries:** "How Many Languages Are There in the World?," Ethnologue, May 3, 2016, https://www.ethnologue.com/guides/how-many-languages.

114 **"needed language development":** "About SIL: Our History," SIL International, May 1, 2012, https://www.sil.org/about/history.

114 **counts 1,300:** Bill Palmer, ed., *The Languages and Linguistics of the New Guinea Area: A Comprehensive Guide*, The World of Linguistics, Vol. 4 (Berlin: De Gruyter Mouton, 2018).

114 **New Guinea had 1,250:** Jared M. Diamond, "The Language Steamrollers," *Nature* 389, no. 6651 (October 1997): 544–46, https://doi.org/10.1038/39184.

115 **One New Guinea isolate is Maybrat:** Palmer, *The Languages and Linguistics of the New Guinea Area*, 1, 9.

116 **a trade language of seafaring Malayans:** R. Nugroho, "The Origins and Development of

Bahasa Indonesia," *PMLA/Publications of the Modern Language Association of America* 72, no. 2 (April 1957): 23–28, https://doi.org/10.2307/2699135.

117 **perhaps as long as 65,000:** Beehler, *New Guinea*, 33–34.

117 **can subsist in isolation:** Daniel Nettle, "Explaining Global Patterns of Language Diversity," *Journal of Anthropological Archaeology* 17, no. 4 (December 1998): 354–74, https://doi.org/10.1006/jaar.1998.0328.

117 **more recent statistical research:** Xia Hua et al., "The Ecological Drivers of Variation in Global Language Diversity," *Nature Communications* 10, no. 1 (December 2019): 2047, https://doi.org/10.1038/s41467-019-09842-2.

118 **185 groups:** Antenor Vaz, *Pueblos Indígenas en Aislamiento: Territorios y desarrollo en la Amazonía y Gran Chaco* [Isolated Indigenous peoples: Territories and development in Amazonia and the Gran Chaco] (New York: Land is Life, 2019).

118 **two largest unbroken expanses of tropical woods:** Matthew C. Hansen et al., "The Fate of Tropical Forest Fragments," *Science Advances* 6, no. 11 (March 2020): eaax8574, https://doi.org/10.1126/sciadv.aax8574.

119 **The Zo'é, for example:** "Quem São Os Zo'é," [Who are the Zo'é?] National Indian Foundation, accessed March 6, 2020, http://www.funai.gov.br/index.php/zoe/2025-quem-sao-os-zo-e.

119 **over 170 million acres:** Hansen et al., "The Fate of Tropical Forest Fragments."

119 **a Jackson Pollack painting:** Dixon and Aikhenvald, *The Amazonian Languages*, 1.

119 **the Amazon's 350 languages:** Alexandra Y. Aikhenvald, *The Languages of the Amazon* (Oxford: Oxford University Press, 2015), 32–59.

120 **championed by Noam Chomsky:** "On Language and Humanity: In Conversation With Noam Chomsky," *The MIT Press Reader* (blog), August 12, 2019, https://thereader.mitpress.mit.edu/noam-chomsky-interview/.

120 **"invariably from a language of Amazonia":** Dixon and Aikhenvald, *The Amazonian Languages*, 1.

120 **new idea of a possessive verb:** Aikhenvald, *The Languages of the Amazon*, 385.

120 **possibilities of possession vary:** Aikhenvald, *The Languages of the Amazon*, 169–70.

121 **retell events in full dialogue:** Aikhenvald, *The Languages of the Amazon*, 250–78.

121 **globally unique sound:** Aikhenvald, *The Languages of the Amazon*, 108; Dixon and Aikhenvald, *The Amazonian Languages*, 354–55.

121 **largest linguistic family with 1,524 entries:** Jonathan Loh and David Harmon, *Biocultural Diversity: Threatened Species, Endangered Languages* (Zeist: WWF–Netherlands, 2014); "What Are the Largest Language Families?," Ethnologue, May 25, 2019, https://www.ethnologue.com/guides/largest-families; Diamond, "The Language Steamrollers."

121 **the Republic of the Congo is home to 60:** "Gabon Languages," Ethnologue, accessed February 21, 2021, https://www.ethnologue.com/country/GA/languages; "Republic of the Congo Languages," Ethnologue, accessed February 21, 2021, https://www.ethnologue.com/country/CG/languages; "Democratic Republic of the Congo," Ethnologue, accessed March 28, 2021, https://www.ethnologue.com/country/cd/languages.

122 **their pre-contact forms of speech:** Bahuchet, "Changing Language, Remaining Pygmy."

122 **reduced by 90 percent:** Fiona Maisels et al., "Devastating Decline of Forest Elephants in

Central Africa," *PLOS ONE* 8, no. 3 (March 4, 2013): e59469, https://doi.org/10.1371/journal.pone.0059469.

122 **long memories:** "Scientists Find Elephant Memories May Hold Key to Survival," EurekAlert!, accessed March 27, 2020, http://www.eurekalert.org/pub_releases/2008-08/wcs-sfe081108.php.

125 **sixty Indigenous languages in thirteen families:** Nick Walker, "Mapping Indigenous Languages in Canada," *Canadian Geographic*, December 15, 2017, https://www.canadiangeographic.ca/article/mapping-indigenous-languages-canada.

125 **Alaska adds another dozen languages:** Gary Holton, Jim Kerr, and Colin West, *Indigenous Peoples and Languages of Alaska*, map (Fairbanks: Alaska Native Language Center and University of Anchorage, 2011), https://www.uaf.edu/anla/collections/map/anlmap.png.

125 **forty "small-numbered" Indigenous groups:** David Nathaniel Berger, *Indigenous World 2019* (Copenhagen: International Working Group for Indigenous Affairs, 2019).

125 **larger Taiga native peoples:** "List of Larger Indigenous Peoples of Russia," in *Wikipedia*, February 22, 2020, https://en.wikipedia.org/w/index.php?title=List_of_larger_indigenous_peoples_of_Russia&oldid=942041685.

129 **William Housty's 2014 paper:** William Housty et al., "Grizzly Bear Monitoring by the Heiltsuk People as a Crucible for First Nation Conservation Practice," *Ecology and Society* 19, no. 2 (June 27, 2014), https://doi.org/10.5751/ES-06668-190270.

129 **Heiltsuk leaders note:** Heiltsuk Tribal Council, "Decision of the Heiltsuk Dáduqvlá Committee Regarding the October 13, 2016 Nathan E. Stewart Spill" (Heiltsuk Tribal Council, 2018), 7.

Chapter 6: Guardians

134 **thirty-six times less likely:** Walker et al., "The Role of Forest Conversion, Degradation, and Disturbance."

138 **Article 231 of Brazil's 1988 constitution:** "Constituição," accessed October 12, 2019, http://www.planalto.gov.br/ccivil_03/constituicao/constituicao.htm.

139 **77 million acres of *resguardos*:** "Resguardos Indigenas en la Amazonía y la Orinoquía" [Indigenous reserves in the Amazon and Orinoco basins], Territorio Indígena y Gobernanza, accessed April 27, 2020, http://territorioindigenaygobernanza.com/web/col_07/.

140 **36 percent of the IFLs:** Fa et al., "Importance of Indigenous Peoples' Lands."

140 **money's allure is more likely:** Kathryn Baragwanath and Ella Bayi, "Collective Property Rights Reduce Deforestation in the Brazilian Amazon," *Proceedings of the National Academy of Sciences of the USA* 117, no. 34 (August 25, 2020): 20495–502, https://doi.org/10.1073/pnas.1917874117.

143 **denied the Haida's existence:** Mark Dowie, *The Haida Gwaii Lesson: A Strategic Playbook for Indigenous Sovereignty* (Oakland, CA: Inkshares, 2017), 114–15.

143 **500 generations:** Dowie, *The Haida Gwaii Lesson*, 93.

143 **Supreme Court surprised the Haida:** Dowie, *The Haida Gwaii Lesson*, 116.

144 **Umbrella Final Agreement:** "Agreements with First Nations," January 17, 2019, https://

yukon.ca/en/agreements-first-nations#modern-treaties-comprehensive-land-claims-agreements.

144 **2012 moratorium on new mining claims:** "Half the Yukon Now Covered by Staking Bans," *Yukon News*, February 3, 2017, https://www.yukon-news.com/news/half-the-yukon-now-covered-by-staking-bans/.

145 **Heiltsuk Declaration of Title and Rights:** Heiltsuk Hemas and Heiltsuk Tribal Council, "Declaration of Heiltsuk Title and Rights," 2015, http://www.heiltsuknation.ca/wp-content/uploads/2015/11/Heiltsuk-Declaration_Final.pdf.

145 **The pipeline was defeated:** Jason Proctor, "Court Overturns Northern Gateway Pipeline Approval," CBC, June 30, 2016, https://www.cbc.ca/news/canada/british-columbia/northern-gateway-pipeline-federal-court-of-appeal-1.3659561.

146 **Congress fast-tracked the ANCSA:** Claus-M. Naske and Herman E. Slotnick, *Alaska: A History*, 3rd ed. (Norman: University of Oklahoma Press, 2011), 283–308.

147 **lucrative contracting ventures:** Michael Grabell and Jennifer LaFleur, "What Are Alaska Native Corporations?," ProPublica, December 15, 2010, https://www.propublica.org/article/what-are-alaska-native-corporations; "Alaska Native Corporations," accessed March 14, 2020, https://www.akrdc.org/alaska-native-corporations.

147 **echo Native customs:** Dixie Dayo and Gary Kofinas, "Institutional Innovation in Less Than Ideal Conditions: Management of Commons by an Alaska Native Village Corporation," *International Journal of the Commons* 4, no. 1 (September 25, 2009): 142, https://doi.org/10.18352/ijc.146.

147 **Native Pete Shaeffer:** Thomas R. Berger, *Village Journey: The Report of the Alaska Native Review Commission* (New York: Hill and Wang, 1985), 95.

148 **Vitus Bering's 1741 reconnaissance voyage:** Naske and Slotnick, *Alaska*, 39–48.

148 **Russian forts were set up:** John F Richards and Ebooks Corporation, *The Unending Frontier: An Environmental History of the Early Modern World* (Berkeley: University of California Press, 2003), 538; Anna Reid, *The Shaman's Coat: A Native History of Siberia* (New York: Walker & Company, 2003), 48–49.

149 **outlawed native languages:** Karl Hele, "Native People and the Socialist State: The Native Populations of Siberia and Their Experience as Part of the the Union of Soviet Socialist Republics," *Canadian Journal of Native Studies* 14, no. 2 (1994): 22.

149 **it has proved fleeting:** Will Englund, "This Russian Group Spoke up for Native People in the Far North. It Was Closed over a Paperwork Error," *Washington Post*, November 17, 2019, accessed January 3, 2021, https://www.washingtonpost.com/world/europe/this-russian-group-spoke-up-for-native-people-in-the-far-north-it-was-closed-over-a-paperwork-error/2019/11/15/a78d0194-061b-11ea-9118-25d6bd37dfb1_story.html; Liubov Sulyandziga and Rodion Sulyandziga, "Russian Federation: Indigenous Peoples and Land Rights," *Fourth World Journal* 20, no. 1 (Summer 2020): 1–19.

150 **separate lands law was enacted:** Sulyandziga and Sulyandziga, "Russian Federation: Indigenous Peoples and Land Rights."

151 **nearly all of the 500-plus TTNUs:** Sulyandziga and Sulyandziga, "Russian Federation: Indigenous Peoples and Land Rights."

153 **rebounded by 50 percent:** Jeremy Hance, "Happy Tigers: Siberian Population Continues

to Grow," Mongabay, June 9, 2015, https://news.mongabay.com/2015/06/happy-tigers-siberian-population-continues-to-grow/.

154 **according to Victoria Tauli-Corpuz:** OHCHR, "End of Mission Statement by the United Nations Special Rapporteur on the Rights of Indigenous Peoples, Victoria Tauli-Corpuz on Her Visit to the Republic of Congo," accessed November 12, 2019, https://www.ohchr.org/EN/NewsEvents/Pages/DisplayNews.aspx?NewsID=25196&LangID=E; Ngambouk Vitalis Pemunta, "Fortress Conservation, Wildlife Legislation and the Baka Pygmies of Southeast Cameroon," *GeoJournal* 84, no. 4 (August 1, 2019): 1035–55, https://doi.org/10.1007/s10708-018-9906-z; Aili Pyhälä, Ana Osuna Orozco, and Simon Counsell, *Protected Areas in the Congo Basin: Failing Both People and Biodiversity?*, Under the Canopy series (London: Rainforest Foundation UK, 2016).

Chapter 7: Forests and the Real Economy

162 **"the nature of economics":** Partha Dasgupta, *The Economics of Biodiversity: The Dasgupta Review* (London: HM Treasury, February 2021), 300.

163 **nine-day walk south to north:** Chase D. Brownstein, "Caesar's Bestiary: Using Classical Accounts to Statistically Map Changes in the Large Mammal Fauna of Germany during the Pleistocene and Holocene," *Historical Biology*, October 10, 2018, 1–10, https://doi.org/10.1080/08912963.2018.1520230.

163 **middle latitudes of Europe:** N. Roberts et al., "Europe's Lost Forests: A Pollen-Based Synthesis for the Last 11,000 Years," *Scientific Reports* 8, no. 1 (December 2018): 716, https://doi.org/10.1038/s41598-017-18646-7.

163 **Rome beavered its way:** John Perlin, *A Forest Journey: The Role of Wood in the Development of Civilization* (New York: W.W. Norton, 1989), 125–27.

164 **40 to 45 percent by 1500:** Roberts et al., "Europe's Lost Forests."

164 **which was almost all of them:** Warren Dean, *With Broadax and Firebrand: The Destruction of the Brazilian Atlantic Forest* (Berkeley: University of California Press, 2008), 45–50.

164 **episodes spanning 500 years:** Dean, *With Broadax and Firebrand.*

164 **less than 8 percent:** L. Patricia C. Morellato and Célio F. B. Haddad, "Introduction: The Brazilian Atlantic Forest," *Biotropica* 32, no. 4b (2000): 786–92, https://doi.org/10.1111/j.1744-7429.2000.tb00618.x.

165 **70 percent of New York State's Adirondack forest:** Jerry Jenkins and Andy Keal, *The Adirondack Atlas: A Geographic Portrait of the Adirondack Park* (Syracuse, NY: Syracuse University Press, Adirondack Museum, 2004), 100–101.

165 **twenty times as much metal:** Perlin, *A Forest Journey*, 337.

165 **Michigan, Ohio, and Wisconsin:** Perlin, *A Forest Journey*, 355.

165 **3.2 million miles:** Douglas W. MacCleery, *American Forests: A History of Resiliency and Recovery*, FS-540 (Washington, DC: United States Department of Agriculture, Forest Service, 1992), 11.

165 **Increase Lapham sent a report:** Increase Allen Lapham, "Report on the Disastrous Effects of the Destruction of Forest Trees, Now Going On So Rapidly in the State of Wis-

consin," 1867, http://digicoll.library.wisc.edu/cgi-bin/WI/WI-idx?type=header;pview=hide;id=WI.RDEFT.

166 **on a par with old growth:** Coeli M. Hoover, William B. Leak, and Brian G. Keel, "Benchmark Carbon Stocks from Old-Growth Forests in Northern New England, USA," *Forest Ecology and Management* 266 (February 15, 2012): 108–114, https://doi.org/10.1016/j.foreco.2011.11.010.

166 **dropped by around 70 percent:** Jonathan R. Thompson et al., "Four Centuries of Change in Northeastern United States Forests," ed. Ben Bond-Lamberty, *PLOS ONE* 8, no. 9 (September 4, 2013): e72540, https://doi.org/10.1371/journal.pone.0072540.

166 **colonists' pigs:** "Landscape History of Central New England," Harvard Forest, accessed September 14, 2019, https://harvardforest.fas.harvard.edu/diorama-series/landscape-history-central-new-england.

166 **310 million acres of forestland:** MacCleery, *American Forests: A History of Resiliency and Recovery*, 3, 11.

167 **"invisible hand":** Adam Smith, *The Wealth of Nations* (Blacksburg, VA: Thrifty Books, 1776), 423.

170 **one-thirtieth as much carbon:** Dryw A. Jones and Kevin L. O'Hara, "Carbon Storage in Young Growth Coast Redwood Stands," in Richard Standiford et al., *Proceedings of Coast Redwood Forests in a Changing California: A Symposium for Scientists and Managers* (Albany, CA: Pacific Southwest Research Station, Forest Service, US Department of Agriculture, January 2012), 515–523; Stephen C. Sillett et al., "Allometric Equations for *Sequoia sempervirens* in Forests of Different Ages," *Forest Ecology and Management* 433 (February 15, 2019): 349–63, https://doi.org/10.1016/j.foreco.2018.11.016.

172 **Not one was environmental:** Ella Koeze, "How the Economy Is Actually Doing, in 9 Charts," *New York Times*, December 17, 2020, sec. Business, https://www.nytimes.com/interactive/2020/12/17/business/economy/economic-indicator-charts-measures.html.

172 **people flocked to nature:** Michael J. Coren and Dan Kopf, "Once Again, a Pandemic Has Stoked Americans' Love for National Parks," Quartz, September 29, 2020, accessed January 4, 2021, https://qz.com/1908674/covid-19-has-americans-visiting-national-parks-in-record-numbers/; Nathan Rott, "'We Had To Get Out': Despite the Risks, Business Is Booming at National Parks," NPR, accessed January 4, 2021, https://www.npr.org/2020/08/11/900270344/we-had-to-get-out-despite-the-risks-business-is-booming-at-national-parks.

173 **cautious ramping up of action:** Jason Shogren and Michael Toman, "How Much Climate Change Is Too Much? An Economics Perspective," Climate Change Issues Brief (Washington, DC: Resources for the Future, 2000); William D Nordhaus, "A Review of the Stern Review on the Economics of Climate Change," *Journal of Economic Literature* 45, no. 3 (September 2007): 686–702.

173 **famous cost-benefit analysis:** "Stern Review on the Economics of Climate Change," Final Report, accessed January 9, 2020, https://webarchive.nationalarchives.gov.uk/+/http://www.hm-treasury.gov.uk/independent_reviews/stern_review_economics_climate_change/stern_review_report.cfm.

173 **more ambiguous conclusions:** Richard S. J. Tol and Gary W. Yohe, "A Review of the

Stern Review," *World Economics* 7, no. 4 (2006): 19; Nordhaus, "A Review of the Stern Review on the Economics of Climate Change."

173 **unrepentant ten years after:** Robin McKie, "Nicholas Stern: Cost of Global Warming 'Is Worse than I Feared,'" *The Guardian*, November 6, 2016, sec. Environment, http://www.theguardian.com/environment/2016/nov/06/nicholas-stern-climate-change-review-10-years-on-interview-decisive-years-humanity.

174 **It debuted in the 1960s:** Binyamin Appelbaum, *The Economists' Hour: False Prophets, Free Markets, and the Fracture of Society* (Boston: Little, Brown and Company, 2019), 185–214, https://www.overdrive.com/search?q=F8A1CECC-5F81-48EE-A676-1741319DBB10.

175 **contract has been broken:** Elinor Ostrom, *Governing the Commons: The Evolution of Institutions for Collective Action*, Canto Classics (Cambridge, UK: Cambridge University Press, 2015), 183–84.

176 **some damage to ecosystems is irreversible:** Dasgupta, *The Economics of Biodiversity: The Dasgupta Review*, 487.

Chapter 8: Money Trees

178 **Publication of the findings:** Charles M. Peters, Alwyn H. Gentry, and Robert O. Mendelsohn, "Valuation of an Amazonian Rainforest," *Nature* 339, no. 6227 (June 1989): 655–56, https://doi.org/10.1038/339655a0.

179 **vulnerability of subsurface northern carbon:** David P. Kreutzweiser, Paul W. Hazlett, and John M. Gunn, "Logging Impacts on the Biogeochemistry of Boreal Forest Soils and Nutrient Export to Aquatic Systems: A Review," *Environmental Reviews* 16, no. NA (December 2008): 157–79, https://doi.org/10.1139/A08-006.

179 **top-ten list of historical emitters:** Kevin A. Baumert, Timothy Herzog, and Jonathan Pershing, *Navigating the Numbers: Greenhouse Gas Data and International Climate Policy* (Washington, DC: World Resources Institute, 2005).

179 **a deal in 1996:** Charles W. Schmidt, "Green Trees for Greenhouse Gases: A Fair Trade-Off?," *Environmental Health Perspectives* 109, no. 3 (2001), http://ezproxy.amherst.edu/login?url=http://search.ebscohost.com/login.aspx?direct=true&db=edsgsc&AN=edsgcl.77276482&site=eds-live&scope=site.

180 **agree on ownership of credits:** Nicole R. Virgilio et al., "Reducing Emissions from Deforestation and Degradation (REDD): A Casebook of On-the-Ground Experience" (Arlington, VA: The Nature Conservancy, Conservation International, Wildlife Conservation Society, 2010), 11, https://www.nature.org/media/climatechange/redd-casebook-tnc-ci-wcs.pdf.

180 **despite being cheaper:** Jonah Busch et al., "Potential for Low-Cost Carbon Dioxide Removal through Tropical Reforestation," *Nature Climate Change* 9, no. 6 (June 2019): 463–66, https://doi.org/10.1038/s41558-019-0485-x.

180 **350 voluntary projects:** Amy E Duchelle et al., "REDD+: Lessons from National and Subnational Implementation," working paper, Ending Tropical Deforestation: A Stock-Take of Progress and Challenges (Washington, DC, 2018).

180 **90 percent of funds:** Seymour and Busch, *Why Forests?*, 365.

181 **"compensated reductions" of deforestation:** Márcio Santilli et al., "Tropical Deforestation and the Kyoto Protocol: An Editorial Essay," in *Tropical Deforestation and Climate Change*, ed. Paulo Moutinho and Stephan Schwartzman (Brasília, DF, Brazil; Washington, DC: Instituto de Pesquisa Ambiental da Amazônia; Environmental Defense Fund, 2005), 47–51.

182 **cleanup costs were 18 to 27 percent:** Robert Stavins et al., "The US Sulphur Dioxide Cap and Trade Programme and Lessons for Climate Policy," *VoxEU.Org* (blog), August 12, 2012, https://voxeu.org/article/lessons-climate-policy-us-sulphur-dioxide-cap-and-trade-programme.

183 **By 2014 it had pledged $4 billion:** Seymour and Busch, *Why Forests?*, 368.

184 **money also moved slowly:** Seymour and Busch, *Why Forests?*, 384–87; Marigold Norman and Smita Nakhooda, "The State of REDD+ Finance," *SSRN Electronic Journal*, 2015, https://doi.org/10.2139/ssrn.2622743.

184 **3 percent of total climate funding:** Climate Focus, *Progress on the New York Declaration on Forests: Finance for Forests—Goals 8 and 9 Assessment Report* (Amsterdam: New York Declaration on Forest Assessment Partners, Climate and Land Use Alliance, 2017), 8, https://www.climatefocus.com/sites/default/files/NYDF%20report%202017%20FINAL.pdf.

185 **representing 30 percent:** Seymour and Busch, *Why Forests?*, 43; Bronson W. Griscom et al., "Natural Climate Solutions," *Proceedings of the National Academy of Sciences of the USA* 114, no. 44 (October 31, 2017): 11645–50, https://doi.org/10.1073/pnas.1710465114.

185 **A couple of academic models:** Bernardo Strassburg et al., "Reducing Emissions from Deforestation—The 'Combined Incentives' Mechanism and Empirical Simulations," *Global Environmental Change* 19, no. 2 (May 2009): 265–78, https://doi.org/10.1016/j.gloenvcha.2008.11.004; Jonah Busch et al., "Comparing Climate and Cost Impacts of Reference Levels for Reducing Emissions from Deforestation," *Environmental Research Letters* 4, no. 4 (October 2009): 044006, https://doi.org/10.1088/1748-9326/4/4/044006.

185 **365 economists surveyed in 2015:** Peter Howard and Derek Sylvan, *Expert Consensus on the Economics of Climate Change* (New York: New York University Institute for Policy Integrity, 2015).

186 **twenty-six countries with higher per capita CO_2:** Hannah Ritchie and Max Roser, "CO_2 and Greenhouse Gas Emissions," Our World in Data, May 11, 2017, https://ourworldindata.org/co2-and-other-greenhouse-gas-emissions.

187 **Gabon and Norway made a REDD+ deal:** "Addendum to Letter of Intent between Gabon and CAFI Signed in 2017," September 22, 2019.

187 **2007 paper (entitled "No Forest Left Behind"):** Gustavo A. B. da Fonseca et al., "No Forest Left Behind," *PLOS Biology* 5, no. 8 (August 14, 2007): e216, https://doi.org/10.1371/journal.pbio.0050216.

190 **75 percent of all overseas support:** David Strelneck and Thaís Vilela, *International Conservation Funding in the Amazon: An Updated Analysis* (Palo Alto, CA: Moore Foundation, Conservation Strategy Fund, 2017), https://www.moore.org/docs/default-source/environmental-conservation/andes-amazon-initiative/international-conservation-funding-in-the-amazon_updated-analysis8eda0461a10f68a58452ff00002785c8.pdf?sfvrsn=d6d56c0c_6.

190 **tax receipts were approximately 3,840 times:** "Brazil's Federal Tax Revenue in 2019 Totals 1.537 Trillion Reais, +1.69% on Year—Tax Agency," *Reuters*, January 23, 2020, https://www.reuters.com/article/brazil-economy-tax-idUSL8N29S4KA; Amazon Fund, *Fundo Amazônia: Relatório de Atividades 2019*, annual report (Brasília, DF, Brazil: Amazon Fund, 2020).

190 **60 percent of respondents:** Cassie Flynn et al., *Peoples' Climate Vote* (New York: UN Development Program and Oxford University, 2021), 39.

191 **MapBiomas:** "Mapbiomas Brasil," accessed July 27, 2020, https://mapbiomas.org/.

Chapter 9: The People's Forest

193 **a 2015 law:** "Russia's Putin Signs Law against 'Undesirable' NGOs," BBC, May 24, 2015, sec. Europe, https://www.bbc.com/news/world-europe-32860526.

193 **That movement is nature conservation:** Douglas R Weiner, *A Little Corner of Freedom: Russian Nature Protection from Stalin to Gorbachev* (Berkeley, CA: University of California Press, 2002), 1–21.

194 **percolating for a couple of decades:** Weiner, *A Little Corner of Freedom*, 28.

194 **28 out of 101 federal zapovedniks:** Vladimir Krever, Mikhail Stishov, and Irina Onufrenya, *National Protected Areas of the Russian Federation: Gap Analysis and Perspective Framework* (Moscow: World Wildlife Fund–Russia, 2009).

195 **pushed these reserves persistently:** Weiner, *A Little Corner of Freedom*, 1–5.

195 **acreage to over 38 million:** "National Parks of Russia," in *Wikipedia*, December 23, 2020, https://en.wikipedia.org/w/index.php?title=National_parks_of_Russia&oldid=995810640.

198 **six categories of strictness:** "Protected Area Categories," IUCN, May 27, 2016, https://www.iucn.org/theme/protected-areas/about/protected-area-categories.

198 **Aldo Leopold argued:** Aldo Leopold, *A Sand County Almanac, and Sketches Here and There* (New York: Oxford University Press, 1949), 201–7.

199 **Congo by between 60,000 and 100,000:** Jerome Lewis, "Forest Hunter-Gatherers and Their World: A Study of the Mbendjele Yaka Pygmies of Congo-Brazzaville and Their Secular and Religious Activities and Representations" (PhD thesis, London School of Economics and Political Science, 2002); Bahuchet, "Changing Language, Remaining Pygmy."

199 **in New Guinea by 50,000:** Beehler, *New Guinea*, 33–34.

199 **in Siberia by 45,000:** Vladimir V. Pitulko et al., "Early Human Presence in the Arctic: Evidence from 45,000-Year-Old Mammoth Remains," *Science* 351, no. 6270 (January 15, 2016): 260–63, https://doi.org/10.1126/science.aad0554.

199 **in North America by 30,000:** Ciprian F. Ardelean et al., "Evidence of Human Occupation in Mexico around the Last Glacial Maximum," *Nature* 584, no. 7819 (August 2020): 87–92, https://doi.org/10.1038/s41586-020-2509-0.

200 **our love of mountain wilderness:** Roderick Frazier Nash, *Wilderness and the American Mind*, 5th ed. (New Haven: Yale University Press, 2014), 96–121.

200 **twenty-six different tribes:** "Associated Tribes—Yellowstone National Park," US National Park Service, accessed July 29, 2020, https://www.nps.gov/yell/learn/historyculture/associatedtribes.htm.

201 **called the Dehcho K'éhodi:** Environment and Climate Change Canada, "First New

Indigenous Protected Area in Canada: Edéhzhíe Protected Area," backgrounders, gcnws, October 11, 2018, https://www.canada.ca/en/environment-climate-change/news/2018/10/first-new-indigenous-protected-area-in-canada-edehzhie-protected-area.html.

202 **called Thaidene Nënë:** Radio Canada International, "New Indigenous Protected Area Created in the Northwest Territories," Radio Canada International, accessed May 12, 2020, https://www.rcinet.ca/en/2019/08/21/new-indigenous-protected-area-thaidene-nene-nwt/.

202 **proposed Ross River Kaska IPCA:** Ross River Dena Council, *An Indigenous Protected Area in the Ross River Dena Territory in the Yukon*, request for funding from the Canada Nature Fund (Ross River, YT, Canada: Ross River Dena Council, August, 2018).

204 **Dutch colonials set aside:** P. Jepson and R. J. Whittaker, "Histories of Protected Areas: Internationalisation of Conservationist Values and Their Adoption in the Netherlands Indies (Indonesia)," *Environment and History* 8, no. 2 (May 1, 2002): 129–72, https://doi.org/10.3197/096734002129342620.

205 **jumped to 26 percent by 2015:** David Morales-Hidalgo, Sonja N. Oswalt, and E. Somanathan, "Status and Trends in Global Primary Forest, Protected Areas, and Areas Designated for Conservation of Biodiversity from the Global Forest Resources Assessment 2015," *Forest Ecology and Management* 352 (September 2015): 68–77, https://doi.org/10.1016/j.foreco.2015.06.011.

205 **46 percent as of 2021:** Ane Alencar et al., *O fogo e o desmatamento em 2019 e o que vem em 2020* [Fire and deforestation in 2019 and what lies ahead in 2020], Nota Técnica no. 3 (Brasília, DF, Brazil: Amazon Environmental Research Institute, April 2020).

205 **Democratic Republic of the Congo, with 60 million:** *Global Forest Resource Assessment 2020*, www.fao.org, accessed August 8, 2020, http://www.fao.org/forest-resources-assessment/2020.

205 **a third as likely to be fragmented:** Potapov et al., "The Last Frontiers of Wilderness."

205 **37 percent of the spectacular drop:** B. Soares-Filho et al., "Role of Brazilian Amazon Protected Areas in Climate Change Mitigation," *Proceedings of the National Academy of Sciences of the USA* 107, no. 24 (June 15, 2010): 10821–26, https://doi.org/10.1073/pnas.0913048107.

205 **reduced forest loss by around 85 percent:** Nolte et al., "Governance Regime and Location Influence."

205 **a statistically significant dent:** Richard Damania and David Wheeler, *Road Improvement and Deforestation in the Congo Basin Countries*, Policy Research Working Papers (Washington, DC: The World Bank, 2015), https://doi.org/10.1596/1813-9450-7274.

206 **30 percent by 2030:** United Nations Convention on Biological Diversity, "Zero Draft of the Post-2020 Global Biodiversity Framework" (United Nations, January 6, 2020).

206 **model developed by Venezuelan biologist:** Janeth Lessmann et al., "Cost-Effective Protection of Biodiversity in the Western Amazon," *Biological Conservation* 235 (July 2019): 250–59, https://doi.org/10.1016/j.biocon.2019.04.022.

206 **$94 billion to feed:** "Global Pet Food Sales 2019," Statista, accessed March 29, 2021, https://www.statista.com/statistics/253953/global-pet-food-sales/.

206 **2019 global GDP:** World Bank, "GDP (Current US$)" data, accessed February 18, 2021, https://data.worldbank.org/indicator/NY.GDP.MKTP.CD.

207 **By one estimate, protecting half:** Judith Schleicher et al., "Protecting Half of the Planet Could Directly Affect Over One Billion People," *Nature Sustainability* 2, no. 12 (December 2019): 1094–96, https://doi.org/10.1038/s41893-019-0423-y.

207 **scientists such as E. O. Wilson:** Edward O. Wilson, *Half-Earth: Our Planet's Fight for Life* (New York: Liveright Publishing, a division of W. W. Norton, 2016).

Chapter 10: Less Roads Traveled

214 **2016 map of the globe's roadless zones:** Pierre L. Ibisch et al., "A Global Map of Roadless Areas and Their Conservation Status," *Science* 354, no. 6318 (December 16, 2016): 1423–27, https://doi.org/10.1126/science.aaf7166; personal communication from Pierre Ibisch and Monika Hoffman with updated roadless area figures (December 31, 2019).

215 **most biologically diverse protected area:** Gorman, "Is This the World's Most Diverse National Park?"

215 **a 1990 scientific expedition in the area:** Theodore A. Parker III and Brent Baily, eds., *A Biological Assessment of the Alto Madidi Region and Adjacent Areas of Northwest Bolivia, May 18 – June 15, 1990*, Rapid Assessment Working Papers (Washington, DC: Conservation International, 1991), 9.

220 **"Westward March":** "Marcha para o Oeste" [Westward march], in *Wikipédia, a enciclopédia livre*, June 5, 2020, https://pt.wikipedia.org/w/index.php?title=Marcha_para_o_Oeste&oldid=58434772.

220 **95 percent of Amazon deforestation:** Christopher P. Barber et al., "Roads, Deforestation, and the Mitigating Effect of Protected Areas in the Amazon," *Biological Conservation* 177 (September 1, 2014): 203–9, https://doi.org/10.1016/j.biocon.2014.07.004.

220 **regional infrastructure program:** "IIRSA 2000–2010," accessed January 9, 2021, http://iirsa.org/en/Page/Detail?menuItemId=28.

221 **most devastating in intact forests:** Damania and Wheeler, *Road Improvement and Deforestation*.

221 **a plague of defaunation:** Laporte et al., "Expansion of Industrial Logging in Central Africa," *Science* 316, no. 5830 (June 2007): 1451, https://science.sciencemag.org/content/316/5830/1451.

221 **roads are like fences:** Stephen Blake et al., "Roadless Wilderness Area Determines Forest Elephant Movements in the Congo Basin," *PLOS ONE* 3, no. 10 (October 28, 2008): e3546, https://doi.org/10.1371/journal.pone.0003546.

221 **skyrockets once you're 15 miles:** Stephen Blake et al., "Forest Elephant Crisis in the Congo Basin," *PLOS Biology* 5, no. 4 (April 3, 2007): e111, https://doi.org/10.1371/journal.pbio.0050111.

223 **have explicit roadless policies:** Chris J. Johnson et al., "Growth-Inducing Infrastructure Represents Transformative yet Ignored Keystone Environmental Decisions," *Conservation Letters* 13, no. 2 (2020): e12696, https://doi.org/10.1111/conl.12696.

223 **In 1924 he created:** Connors, *Fire Season*, 128.

224 **14 million acres of the national forests:** Nash, *Wilderness and the American Mind*, 206.

224 **group that was growing fast:** "Special Areas; Roadless Area Conservation," Federal Reg-

ister, January 12, 2001, https://www.federalregister.gov/documents/2001/01/12/01-726/special-areas-roadless-area-conservation.

224 **20 percent of needed maintenance money:** "Special Areas; Roadless Area Conservation."

225 **67 million acres in Alaska:** James Strittholt et al., *Mapping Undisturbed Landscapes in Alaska*, Overview Report (Washington, DC: Conservation Biology Institute, Global Forest Watch, World Resources Institute, 2006), https://wriorg.s3.amazonaws.com/s3fs-public/pdf/gfw_alaska_final.pdf.

225 **2018 *New York Times* op-ed:** Mike Dombeck, "Turning Back the Clock on Protecting Alaska's Wild Lands," *New York Times*, March 13, 2018, sec. Opinion, https://www.nytimes.com/2018/03/13/opinion/protecting-alaska-wild-lands.html.

226 **new jungle access route cost:** "Cotapata National Park and Integrated Management Natural Area," accessed May 27, 2021, https://www.parkswatch.org/parkprofile.php?l=eng&country=bol&park=conp&page=thr.

226 **a 1999 study:** John Reid, *Two Roads and a Lake: An Economic Analysis of Infrastructure Development in the Beni River Watershed / Dos Caminos y Un Lago: Análisis Económico Del Desarrollo de Infraestructura En La Cuenca Del Río Beni* (Philo, CA: Conservation Strategy Fund, 1999), https://www.conservation-strategy.org/en/publication/two-roads-and-lake-economic-analysis-infrastructure-development-beni-river-watersheddos-#.XZ-uDyV7mL9.

226 **affirmed by another report:** Leonardo C. Fleck, Lilian Painter, and Marcos Amend, *Carreteras y áreas protegidas: Un análisis económico integrado de proyectos en el norte de la Amazonía Boliviana* [Roads and protected areas: An integrated economic analysis of projects in the northern Bolivian Amazon], Technical Series no. 12 (La Paz, Bolivia: Conservation Strategy Fund, 2007), 51, https://www.conservation-strategy.org/sites/default/files/field-file/12_Fleck_Carreteras_Norte_Bolivia.pdf.

227 **$788 million:** Matthew M. Taylor, "The Odebrecht Settlement and the Costs of Corruption," (blog), Council on Foreign Relations, accessed August 24, 2020, https://www.cfr.org/blog/odebrecht-settlement-and-costs-corruption.

229 **delivering 77 percent:** Thais Vilela et al., "A Better Amazon Road Network for People and the Environment," *Proceedings of the National Academy of Sciences of the USA* 117, no. 13 (March 2020): 7095–7102, https://doi.org/10.1073/pnas.1910853117.

230 **2015 World Bank study on the Congo:** Damania and Wheeler, *Road Improvement and Deforestation*; Alamgir et al., "Infrastructure Expansion Challenges Sustainable Development in Papua New Guinea."

230 **lies outside of the wild forests:** William F. Laurance et al., "A Global Strategy for Road Building," *Nature* 513, no. 7517 (September 2014): 229–32, https://doi.org/10.1038/nature13717.

230 **a system called Tremarctos:** "Tremarctos Colombia," accessed May 24, 2020, http://www.tremarctoscolombia.org/.

231 **Fleck calculated:** Leonardo C. Fleck, *Eficiência Econômica, Riscos e Custos Ambientais da Reconstrução da Rodovia BR-319* [Economic efficiency, risks and environmental costs of rebuilding the BR-319 highway], Technical Series no. 17 (Lagoa Santa, MG, Brazil: Conservation Strategy Fund, 2009), 55–59.

232 **a story about one logger:** Slaght, *Owls of the Eastern Ice*, 134–36.

233 **roadless parts of Canada:** Government of Canada, Indigenous and Northern Affairs Canada, "How Nutrition North Canada Works," organizational description; promotional material, November 9, 2014, https://www.nutritionnorthcanada.gc.ca/eng/1415538638170/1415538670874.

233 **cities in the Brazilian Amazon:** "Subsídios restritos à Amazônia Legal—Negócios—Diário do Nordeste" [Subsidies restricted to the legal Amazon—business section—Diário do Nordeste], accessed August 24, 2020, https://diariodonordeste.verdesmares.com.br/negocios/subsidios-restritos-a-amazonia-legal-1.1699311.

Chapter 11: Making Nature

236 **cap up to 10 feet:** Richard Hansen, "The Beginning of the End: Conspicuous Consumption and Environmental Impact of the Preclassic Lowland Maya," in *An Archaeological Legacy: Essays in Honor of Ray T. Matheny*, ed. Deanne G. Matheny, Joel C. Janetski, and Glenna Nielsen (Provo, UT: Museum of Peoples and Cultures, Brigham Young University, 2012), 263.

236 **seven to ten times as much:** Busch et al., "Potential for Low-Cost Carbon Dioxide Removal through Tropical Reforestation."

237 **900-plus billion tons:** Karl-Heinz Erb et al., "Unexpectedly Large Impact of Forest Management and Grazing on Global Vegetation Biomass," *Nature* 553, no. 7686 (January 2018): 73–76, https://doi.org/10.1038/nature25138.

237 **The Bonn Challenge:** "The Challenge," Bonn Challenge, accessed March 29, 2020, https://www.bonnchallenge.org/content/challenge.

237 **They had delivered 56 percent:** Radhika Dave et al., *Second Bonn Challenge Progress Report: Application of the Barometer in 2018* (Gland, Switzerland: International Union for Conservation of Nature, 2019), https://doi.org/10.2305/IUCN.CH.2019.06.en.

238 **40 to 80 percent of nitrogen:** T. E. Reimchen et al., "Isotopic Evidence for Enrichment of Salmon-Derived Nutrients in Vegetation, Soil, and Insects in Riparian Zones in Coastal British Columbia," paper presented at the American Fisheries Society Symposium, 2002, 12; Anna Kusmer, "There's Something Fishy about These Trees . . . | Deep Look," KQED, accessed January 5, 2021, https://www.kqed.org/science/1915421/theres-something-fishy-about-these-trees-deep-look.

238 **45 percent of participating countries' plans:** Simon L. Lewis et al., "Restoring Natural Forests Is the Best Way to Remove Atmospheric Carbon," *Nature* 568, no. 7750 (April 2019): 25–28, https://doi.org/10.1038/d41586-019-01026-8.

240 **A 2019 study employed three criteria:** Pedro H. S. Brancalion et al., "Global Restoration Opportunities in Tropical Rainforest Landscapes," *Science Advances* 5, no. 7 (July 2019): eaav3223, https://doi.org/10.1126/sciadv.aav3223.

240 **25 to 37 million acres:** Cláudio Aparecido de Almeida et al., "High Spatial Resolution Land Use and Land Cover Mapping of the Brazilian Legal Amazon in 2008 Using Landsat-5/TM and MODIS Data," *Acta Amazonica* 46, no. 3 (September 2016): 291–302, https://doi.org/10.1590/1809-4392201505504.

241 **Brazil lose 220 million acres:** IG Último Segundo (with information from the Globo

Agency), "Na contramão do país, abertura de pastagens cresce na Amazônia, diz MapBiomas," [Headed in the opposite direction from the rest of the country, clearing for pastures increases in the Amazon according to MapBiomas], Último Segundo, August 29, 2019, https://ultimosegundo.ig.com.br/ciencia/meioambiente/2019-08-29/na-contramao-do-pais-abertura-de-pastagens-cresce-na-amazonia-diz-mapbiomas.html.

241 **290 pounds per square inch:** P. P. J. van der Tol et al., "The Vertical Ground Reaction Force and the Pressure Distribution on the Claws of Dairy Cows While Walking on a Flat Substrate," *Journal of Dairy Science* 86, no. 9 (September 2003): 2875–83, https://doi.org/10.3168/jds.S0022-0302(03)73884-3.

242 **8 pounds per square inch:** "Ground Pressure," in *Wikipedia*, December 19, 2019, https://en.wikipedia.org/w/index.php?title=Ground_pressure&oldid=931493617.

242 **nearly $1,000 per acre:** Roberto Kishinami et al., *Quanto o Brasil precisa investir para recuperar 12 milhões de hectares de floresta?* [How much does Brazil need to invest to restore 12 million hectares of forest?] (São Paulo, SP, Brazil: Instituto Escolhas, 2016), http://www.escolhas.org/wp-content/uploads/2016/09/quanto-o-brasil-precisa.pdf.

242 **1 percent in a normal year:** Paulo Monteiro Brando et al., "Abrupt Increases in Amazonian Tree Mortality Due to Drought–Fire Interactions," *Proceedings of the National Academy of Sciences of the USA* 111, no. 17 (April 29, 2014): 6347–52, https://doi.org/10.1073/pnas.1305499111.

243 **99 out of every 100 acres:** "A Complex Prairie Ecosystem—Tallgrass Prairie National Preserve (US National Park Service)," 2018, https://www.nps.gov/tapr/learn/nature/a-complex-prairie-ecosystem.htm.

244 **undesignated intact forests:** Claudia Azevedo-Ramos and Paulo Moutinho, "No Man's Land in the Brazilian Amazon: Could Undesignated Public Forests Slow Amazon Deforestation?," *Land Use Policy* 73 (April 2018): 125–27, https://doi.org/10.1016/j.landusepol.2018.01.005.

244 **350 million tons of carbon:** "Undesirable Russian Forest," accessed March 29, 2020, https://maps.greenpeace.org/maps/aal_story/.

245 **100 million of the trees:** Lisa Friedman, "A Trillion Trees: How One Idea Triumphed Over Trump's Climate Denialism," *New York Times*, February 12, 2020, sec. Climate, https://www.nytimes.com/2020/02/12/climate/trump-trees-climate-change.html.

245 **Davos website fizzed:** "One Trillion Trees—Uniting the World to Save Forests and Climate," World Economic Forum, accessed April 11, 2020, https://www.weforum.org/agenda/2020/01/one-trillion-trees-world-economic-forum-launches-plan-to-help-nature-and-the-climate/.

246 **there's enough space:** Jean-Francois Bastin et al., "The Global Tree Restoration Potential," *Science* 365, no. 6448 (July 5, 2019): 76–79, https://doi.org/10.1126/science.aax0848.

246 **roughly continues to the present:** "Global Forest Resource Assessment 2020," xii, 15. http://www.fao.org/documents/card/en/c/ca9825en/; Food and Agriculture Organization of the United Nations, *Global Forest Resources Assessment 2015*, Desk Reference (Rome, 2015), 3. http://www.fao.org/3/i4808e/i4808e.pdf.

246 **"a distraction":** Theodore Schleifer, "Marc Benioff Picks a New Fight with Silicon Valley—Over Trees," Vox, January 21, 2020, https://www.vox.com/recode/2020/1/21/21075804/marc-benioff-trees-silicon-valley-donald-trump-davos.

Chapter 12: An Invitation

249 **"success of a hopeless cause!":** Fred Strebeigh, "Lenin's Eco-Warriors," *New York Times*, August 7, 2017, sec. Opinion, https://www.nytimes.com/2017/08/07/opinion/lenin-environment-siberia.html.

250 **cancel dozens of parks in a day:** Weiner, *A Little Corner of Freedom*, 104–36.

253 **trigger dopamine:** Kazuhiro Sumitomo et al., "Conifer-Derived Monoterpenes and Forest Walking," *Mass Spectrometry* 4, no. 1 (2015): A0042 , https://doi.org/10.5702/massspectrometry.A0042.

INDEX